WITHDRAWN

Carnegie Mellon

SNOW IN THE CITIES

SLEIGHING IN CENTRAL PARK
(*Harpers*, 1890)
Courtesy of New York State Historical Association, Cooperstown. An elite version of a pastime many residents of northern cities had enjoyed for over 200 years and some revive a century later.

SNOW IN THE CITIES

A History of America's Urban Response

by
BLAKE MCKELVEY

UNIVERSITY OF ROCHESTER PRESS

Copyright © 1995 University of Rochester Press

All Rights Reserved. Except as permitted under current legislation, no part of this work may be photocopied, stored in a retrieval system, published, performed in public, adapted, broadcast, transmitted, recorded or reproduced in any form or by any means, without the prior permission of the copyright owner.

First published 1995

University of Rochester Press
34–36 Administration Building, University of Rochester
Rochester, New York, 14627, USA
and at PO Box 9, Woodbridge, Suffolk IP12 3DF, UK

ISBN 1 878822 54 3

Library of Congress Cataloging-in-Publication Data

McKelvey, Blake, 1903–
    Snow in the cities : a history of America's urban response / by Blake McKelvey.
    p. cm.
    Includes bibliographical references and index.
    ISBN 1-878822-54-3 (cloth : alk. paper)
    1. Snow removal—United States—History. 2. Snow—United States—History. I. Title.
TD868.M38 1995
388.3' 12—dc20   95–16010

British Library Cataloguing-in-Publication Data

McKelvey, Blake
    Snow in the Cities: History of America's Urban Response
    I. Title
    551.5550973
    ISBN 1-878822-54-3

This publication is printed on acid-free paper

Printed in the United States of America

CONTENTS

Preface		vii
1	Snowstorms in Pedestrian Towns	1
2	Sleighbells and Steam Whistles in the Snow	19
3	Storm Warnings, Snow Plows and Blizzards	41
4	From Snow-Plowing to Snow Removal	67
5	Snow-Fighting in Motorized Cities	99
6	Snow-Fighting in a Metropolitan Area	129
7	Urban Snow Problems Widely Shared	157
Snowfall in American Cities – map and table		194
Index		197

PREFACE

This book is a product as well as a record of snowstorm experiences. In the first place, as an author who has long since passed the proverbial age of three score and ten, all of it in snow country, I have many recollections of harsh but exciting bouts with heavy snows. I have also enjoyed exchanging tales of such encounters with friends who boasted even more dramatic ordeals. Indeed, the incentive to write a book about urban snowstorms came some twenty years ago from the first-hand experience of a correspondent in a distant city. After clambering through unusually heavy drifts in New Brunswick, New Jersey, to reach his office, my friend introduced his reply to a query about other matters with a sprightly touch:

> Your letter of Jan. 10 has arrived, presumably conveyed hither by dog sled or airlift. Don't you think it would make an interesting short book to describe historically what cities have attempted to do in the past, and in the more fantastic present, with heavy snowfalls?

To a scholar residing in Rochester and long experienced with the rigors of its winters, such a suggestion was a challenge. As its official City Historian, I promptly seized the opportunity to write a twenty-four page article on "Snowstorms and Snow Fighting: The Rochester Experience." In its preparation I discovered not only that Rochester was a reluctant contender for the title of "Snow Capital of America," but also that the topic had many fascinating civic and technological ramifications. I found, too, that almost everybody was interested in the subject. When my article appeared, I received numerous requests for a talk on snowstorms, and in every case, members of the audience had experiences of their own to relate.

In the preparation of that paper I consulted with local weather bureau and public works officials and, to broaden my perspective, with similar authorities in neighboring Buffalo and Syracuse. In occasional interviews, busy people seemed glad to lean back and reminisce over some of the harsh snowstorms they had experienced. Clearly it was a topic of wide interest. Other responsibilities intervened, but my file of notes on snowstorms expanded, and the time to use them has now arrived.

This book adheres fairly closely to the original suggestion. It is an historical review of urban snowstorm experience and response. It does not attempt to tell city administrators how to combat storms; that task has frequently been assigned to engineering firms and other consulting experts. Where possible, I have taken note of their findings as well as of the views exchanged by snow-fighting officials at recurrent conferences on the subject. But the urban response to snowstorms began many decades before municipal officials became aware of the severity of the problem. For two full centuries, in fact, private and occasionally cooperative efforts sufficed as residents of colonial cities adjusted to the rigors of their wintertime environment. Only with the development in the 1800s of agencies for communication and commerce did concerted effort to maintain services appear. Official records are of course lacking for these early eras, but scattered private and journalistic accounts help to reconstruct the story of the varied accommodations of pre-industrial cities to harsh wintertime blasts. And among these accounts are frequent references to light-hearted citizen responses to heavy snowstorms. Indeed, if the early cities set an example worth noting, it was in their capacity to appreciate and find delight in a snow covered environment.

Preface

Just as this book is not a "how-to" manual for administrators, it is also not a scholarly contribution to meteorology. As my research progresses, however, I discovered David M. Ludlum's fascinating volumes on *Early American Winters*. His carefully documented report on the weather records of colonial and early American cities prompted me to push my study back to its earliest beginnings. I became so indebted to him that I asked him to read my first two chapters and advise me candidly whether he felt that I had poached unduly on his researches. Mr. Ludlum replied promptly, declaring that snow was his favorite hobby and that he would welcome a scholarly use of his work. He soon dispatched additional materials, including his painstakingly compiled *Weather Record Book: The Outstanding Events 1871–1970*. He has read an early draft of all seven chapters and supplied numerous helpful suggestions. Finally he has permitted me to include his name on my title page as "Consultant." Readers of his books and of *Weatherwise*, which he has produced and edited since 1948, will realize how inadequately the term "consultant" describes his contributions.

My emphasis, however, is on the urban response, which changed remarkably as the man-made urban environment developed and became more proficient in its services and more vulnerable to obstructions. Mounting hazards brought new efforts to combat them and called for increasing budgetary outlays. The response in most instances was related to the severity of the snow problem, and the official meteorological records supply the standard measurement. Even these abundant records fail, however, to describe the problem in many parts of a large city, and the local response to conditions in the streets was often more graphically portrayed by officials and reporters "on the beat." I have restrained my use of such terms as "record-breaking" to only those storms supported by official records, though of course a truly record-breaking storm at any "point in time" may be outclassed by another of later date. And a twenty-inch storm, as reported by a snow-fighting superintendent, may be recorded by the local weather station at the airport on the outskirts of a city as somewhat more or less. The detailed measurements I have used are descriptive, not definitive, but with Mr. Ludlum's assistance I have tried to represent as accurate a picture as the combined records permit.

A further caution is perhaps desirable. Hasty readers may gain the impression that some cities, particularly New York, have been fighting snow interminably. In a span of 350 years there are a lot of winters, and if you find thirty or so snowy seasons and describe them in some detail it may appear that snowy weather was the dominant aspect of the environment. A few cities in the snowbelt have had winters, snowy winters, ever year, but even there spring also arrives when, as one early clergyman put it, quoting Solomon, "the Voice of the Turtle is heard in the Land." As for New York, most frequently mentioned in these pages, even a light snow presented problems in its tumultuous streets and demanded the most deliberate response. Moreover, the availability of a meticulous index to the *New York Times* and of film copies of that paper over the last hundred years has facilitated a close study of its snowstorm experience, comparatively light though it has been in most winters.

In my search for details of the local response elsewhere, I have again relied primarily on contemporary newspaper reports. Numerous librarians, as listed in my acknowledgements, have cooperated by supplying print-outs of articles on specific storms to which I had found leads in varied indexes; some obligingly added copies of other storm clippings they had assembled. Their interest in the topic, like that of many other officials and individuals consulted, has been encouraging and has enabled me to compile a mass of detail, which, considered in connection with the size and technological stage of the cities involved, supports, I believe, my major thesis: that the interaction of

Preface

snowstorm experiences and responses in successive periods has played a significant role in the evolution of the American urban society. Moreover, the dramatic character of a harsh snowstorm, as has often been noted, has both fostered a sense of community and stimulated a memory of its past.

Little did I dream, when I started to write this book in December 1975, that the next two Decembers would usher in a pair of truly landmark winters—1976–77, the coldest in recorded experience; 1977–78, unsurpassed in the ferocity and breadth of its ravages. Shivering in frigid Rochester in January 1977, I watched from a safe distance as nearby Buffalo battled a succession of blizzards that shattered all metropolitan snowstorm records. And a year later, between forays through record January drifts at home, I watched video reports of more unusual blizzard conditions in Louisville, Kentucky and a dozen other cities that seldom experience harsh winters. While penning my final pages, I have been able to take note of the breakthroughs these hardships have produced in snow-fighting programs both locally and nationally. Seldom have events provided an historian with a more dramatic concluding crescendo both of nature's awesome forces and of man's innovative responses.

And now, almost two decades later and after several more "Remarkable Providences," as Increase Mather christened the major storms of the Colonial period, my updated draft, not to be outdone, finds that the Great Blizzard of 1993 not only surpasses all of its predecessors but provides an instructive climax to America's snowstorm experience and response.

INTRODUCTION

The growth of cities has transformed one of nature's glorious wonders into a perplexing and costly problem. The frosty white blankets that have repeatedly shrouded the landscape in northern states each winter, delighting youngsters and challenging their elders in rural villages since colonial days, have presented increasing hazards to our more complex metropolitan society. Yet many city-dwellers have continued to find inspiration and enchantment in snowstorms. The heightened sense of community brought by a big storm, and the awe of overpowering nature, both endure. So does our zest for winter sports. Watching the weather, battling the weather, predicting its changes, and trying to determine them are integral aspects of urban culture. Coping with snow has also become a recurrent task for urban officials, at least in northern cities.

Whether Mayor John V. Lindsay of New York had any real prospects of election to the White House is problematic. But certainly the popular speculation concerning his chances dropped off sharply after the crippling snowstorm of February 1969 paralyzed extensive parts of the metropolis. The widely published photos of Mayor Lindsay wading through the drifted streets of Queens to demonstrate his concern broadcast an image of failure that was hard to shake off.[1]

Snowstorms have seldom if ever created political heroes. Urban administrators have generally considered themselves lucky to escape criticism after a severe storm. Too

[1] John V. Lindsay, *The City* (New York: Norton, 1970), p. 23.

frequently they have heard their opponents chant, as at Rochester in 1945, "In November remember December."[2]

Yet most snowstorms have produced heroes—modest citizens who rose to unexpected challenges in memorable ways. Such a man was C. W. Wicks, a milk huckster in Brooklyn who, while making his round at 3 A.M. on the first morning of the great blizzard of 1888, encountered a wholesaler's truck loaded with eighty-two large cans of milk stalled in a drift. Realizing that milk would soon be in short supply, he enlisted the aid of neighbors in transporting the cans to Dr. Shena's Sanitarium for mothers with infant babies, making it a haven for sick and needy children in that congested district. And, inspired by the fury of the storm, which cost him a frozen ear, he refused to exact a profit—a decision he later cherished.

Many inhabitants of New York in that blizzard long remembered it as "the most vivid experience of my life." Like the snow-bound of other blizzards in New York and elsewhere, they regarded themselves as heroes just for surviving. Some would recall its hardships and the details of their experience in a revealing record compiled and published fifty years later by "The Blizzard Men of 1888." That document[3], with its recollections of the activities of several hundred individuals, chiefly in humble stations, is of greater interest to the social historian than to the student of meteorology. It convincingly demonstrates, as we shall see, the neglected value of snowstorm reminiscences to students of community development.

Few records have been as long and as faithfully kept as those of local weather observers. None are as avidly and widely followed today as those of the meteorologists. It is somewhat surprising, therefore, to discover that few of the

[2] Blake McKelvey, *Rochester: an emerging metropolis, 1925–1961* (Rochester, N.Y.: Christopher Press, 1961), p. 181.
[3] Samuel M. Strong, ed., *The Great Blizzard of 1888* (New York: 1939). The "Blizzard Men of 1888" met annualy. Samuel Strong gathered their reminiscences and printed them as a memorial pamphlet.

Introduction

increasing number of urban historians have consulted these records. As one of the lonely exceptions, I must concede that possibly the weather had a lesser impact on the history of most cities than it has had on Rochester, which is located in one of America's heaviest snowbelts, but the total neglect of the subject has left a gap in the urban story.

The snowstorm experiences and responses of American cities supply fruitful material for environmental as well as urban historians. As with most historical subjects, the topic is multi-dimensional. Not only was the natural environment pulsating and unpredictable, buffeting selected cities at different times and with varied frequency and intensity, but the impact of similar storms also changed radically as the cities grew in size and technological complexity. And as the man-made urban environment developed, its response expanded to include, in addition to private and cooperative adjustments, concerted efforts by agencies of commerce and communication to maintain their services; in time it also inspired municipal regulations that led finally to active civic programs for snow plowing and snow removal. The early innovations, appearing as spasmodic responses in hard-pressed cities, generally acquired wider application in the wake of excessively harsh winters; only in recent decades, as the urban economy has become completely motorized, have the hazards even of moderate storms gained wide recognition, alerting cities and states throughout the snow country, and the federal government as well, to new wintertime responsibilities.

The urban response to snowstorms falls into a succession of historic eras. Each era had its special technological characteristics and gave way to a new era as technological breakthroughs occurred. Each era had its particular economic and social patterns, and these changed as the structure of urban society developed and brought new community functions. Yet in spite of their many remarkable improvements, the cities faced a continuing challenge in communications and supply. And if the frigid onslaught of successive winters with their varied and still unpredictable deposits of snow never held more than a temporary and slippery grasp on the throttle of urban growth, they supplied repeated reminders to city dwellers that, despite improved facilities, they remained residents of a real world and subject to the as yet inscrutable whims and the untamed forces of nature. Inhabitants of a dozen teeming cities in the Northeast learned that lesson anew in the "standout" winter of 1976–77.

Residents of the solely "pedestrian" towns and cities of the first two centuries of American history held few doubts concerning their subjectivity to nature's forces. Many regarded the wintry blasts as an expression of God's wrath and made only humble efforts to accommodate to these "Remarkable Providences," as Increase Mather termed them.[4] At the start, city-dwellers possessed few implements to cope with storms. Their equipment consisted of wooden shovels, crude sleds, constructed sleighs, and imported thermometers and barometers to supplement the early yardsticks in recording the impact of storms. Best of all, they learned to use the more moderate falls of snow as an aid in transporting heavy loads of produce and other supplies to and from outlying settlements.

Scattered efforts to open snow-drifted roads by driving a battery of horse-drawn sleds through the principal streets heralded the opening of a second era—that of the "horse-and-buggy" city. Of course, light buggies were kept in the barn during snowstorms, even in bustling cities, as the first "horseless carriages" would later be stored. Light sleighs or "cutters" seasonally replaced them, adding a new delight to winter travel. Sleigh riding became popular in northern cities, and racing cutters acquired priority on favored avenues.[5] The early horse-car lines frequently

---

[4] Increase Mather, *Remarkable Providences* (London: J. R. Smith, 1856), cited by Ludlum, op. cit., p. 16.

[5] Blake McKelvey, "Snowstorms and Snow Fighting: The Rochester Experience," *Rochester History* XXVII (Jan., 1965): pp. 6, 8.

Introduction

shifted to sleighs too, when a heavy snow buried the tracks, but their charters generally called for a clearing of the tracks, and this requirement brought the first horse-drawn plows onto the scene in several cities. Conflicts between the managers of the lines endeavoring to maintain their service and the owners of the sleighs and horse sleds, who protested the uneven mounds of snow pushed into the roadway, provided a foretaste of other controversies destined to arise as cities struggled to meet the increasingly complex snow problems.

The great blizzard of 1888 dramatically highlighted the increasing complexity of the annual snow battles that confronted many cities. Improvements in recording, communicating, and analyzing snowstorm records, greatly helped by the development of the telegraph system, had fostered the growth of a science of meteorology that was able by the mid-eighties to provide storm warnings of some value. The even more dramatic development of the nation's railroads had speeded the growth of cities but had rendered them increasingly dependent on the maintenance of service especially in the supply of fuel and provisions. Railroads had pioneered the introduction of snowplows, first with horses, then as attachments to cars pushed by steam engines. But when the great blizzard of 1888 blockaded main and feeder lines throughout the Northeast, essential food and fuel supplies were shut off. Snow and ice, coupled with raging winds, toppled not only the telegraph lines but the newly introduced telephone wires, effectively halting all communications. Only the undersea cables from New York and Boston to London enabled dispatchers in the two ports to exchange news of their calamities and knowledge of developments elsewhere as it filtered into either city. Thus the recorded sufferings of New York in the great blizzard of 1888 provide a graphic view of urban developments and limitations under harsh winter siege.[6]

[6] Strong, *The Great Blizzard of 1888*.

Snow in the Cities

The almost total collapse of services was of course only temporary, and New York with other cities pressed forward with the development of urban technological improvements. A rapid extension of telephone services courted new crises because of fallen wires and prompted New York and a few other storm-threatened cities to bury them in underground conduits. The replacement of horse-cars by trollies made the acquisition of snowplows more imperative, and their wide use challenged the efforts of horse-drawn sleds to maintain a smooth snow road. This contest led inevitably to a recognition of the need to cart some of the snow to convenient dumps and led to disputes over responsibility for the cost. The heavy losses suffered by urban business leaders from periodic snow blockades proved sufficient in one city or another to prompt officials to expand their efforts to plow the streets and in a few cases to remove the snow from crucial intersections and districts. New York City, despite its comparatively moderate snowfalls, suffered most and took the lead in all forms of snow fighting.[7]

The introduction and multiplication of automobiles after the turn of the century precipitated these efforts in some cities, notably Rochester, which had a comparatively large number of motor cars as well as heavy deposits of snow. But again it was a widespread blizzard in February 1914 that prompted the officials of snowbound cities to collaborate in a search for improved snow-fighting techniques. When one distraught street superintendent asked why somebody could not devise a machine to replace the plodding snow shoveler and cart, a manufacturer took no time to announce that he had one ready for use.[8] Soon the pages of *The Municipal Journal*, *The American City*, and other magazines were sprinkled with ads announcing new

[7] George E. Waring, Jr. *Street Cleaning* (New York: Doubleday & Maclure, 1898), pp. 94–99.

[8] McKelvey, op. cit., pp. 11–13; *Scientific American*, and Mar. 14; Apr. 4, 1914; ibid., *Supplement* Mar. 6, 1915.

snow-plows, snow-loaders, and other devices. Articles reporting the efforts of scattered cities to combat heavy snows with the new equipment multiplied, and street superintendents vied in drafting strategies to meet varied snow emergencies.

As the mechanized battlefront spread throughout northern cities, a few began to supplement plowing with sanding and some by spreading salt. The chemical attack stirred a heated debate over pollution and related damages and further complicated the annual wintertime siege. Mounting costs focused new attention on forecasting and on other efforts to prepare for and combat the periodic snowstorms. Hard pressed cities turned increasingly to state and federal agencies for assistance, in this as in other civic fields, as we shall see in chapters VI and VII.

Meanwhile, many residents of the cities continued to derive pleasure and excitement from their winter ordeals. Horse sleighing disappeared, but other winter sports mushroomed, most notably skiing, which gave birth to numerous winter resorts and spawned a multitude of ski trails and slides in the vicinity of all cities in the snowbelt. Some industrialists assisted by supplying lifts to carry impatient skiers up the slopes; others produced snow-making machines to keep the slopes in shape; still others developed snowmobiles to provide speed-hungry customers with horseless sleds. But in spite of all urbanized efforts to predict, combat, exploit, and otherwise cope with heavy winter snows, the recurrent blizzards that sweep over many northern cities *still* possess awe-inspiring force and often leave a glistening landscape that can briefly inspire snow-bound residents with a renewed sense of community with their neighbors and a revived affinity with nature.

They also prompt reminiscence of earlier storms. As John F. Watson wrote in his *Annals of Philadelphia* almost a century and a half ago, "An old fashioned snowstorm, such as we had lately on the 20th and 21st of February, 1829, is the best thing in our country to bring to recollection olden times."⁹

Acknowledgments

In addition to my considerable indebtedness to David M. Ludlum, as noted in the Preface, I owe a special expression of thanks and appreciation to several members of the staff of the Rochester Public Library who have assisted me in locating elusory materials. Dr. Harold Hacker, Director of Libraries, Mrs. Dorothy Humes and Miss Marion McGuire, head and assistant head of the Central Library, Mr. Robert W. Eames, head of the Reference Division, Mrs. Elizabeth Chase, head of interlibrary loans, Miss Judith Prevratil, head of the Division of Science and Technology, and Mr. Wayne Arnold and Miss Shirley Iverson, head and assistant in the Local History Division have each been very helpful and tolerant of the many requests of their former colleague. Mr. Ben Bowman, formerly director of the Rush Rhees Library of the University of Rochester, and his assistants Mr. H. Bradford Smith and Miss Theodora Mills have made their resources conveniently available.

I am particularly indebted to several individuals in Rochester for consultation on aspects of this study. City Manager Elisha Freedman, Commissioner of Public Works Edward F. Watson, and Mr. John D. Hostutler, general secretary of the Industrial Management Council, have each granted interviews and supplied documents related to Rochester's snow-fighting procedures. Mr. W. Stephen Thomas, director-emeritus of the Rochester Museum and Science Center, and Mr. Holman J. Swinney,

---

⁹ John F. Watson, *Annals of Philadelphia & Pennsylvania in the Olden Times* vol. 2 (Philadelphia: E. S. Stuart, 1898), p. 347.

Introduction

director of the Margaret Woodbury Strong Museum, have each made helpful suggestions of available sources. Dr. Joseph W. Barnes, my successor as City Historian, has kindly read the manuscript and made numerous helpful suggestions. His and my former secretary, Mrs. Faye Geldin, has generously supplied assistance beyond the call of duty to this project.

But my efforts to assemble the records of the response of many cities to their snowstorm experiences would have been frustrated without the cooperation of many librarians in widely scattered institutions. The following listing is an inadequate recognition of the contributions I have received:

Adirondack Museum, William K. Verner, curator
American Public Works Association, Donald M. Fairile, director
Buffalo & Erie County Public Library, M.s Ruth Willet, History Department
Buffalo Government Research Bureau, Charles R. Dawson, director
Buffalo Historical Society, Dr. Lester Smith
Carnegie Library of Pittsburgh, Mrs. Lucille A. Tomko, Pennsylvania Division
Cleveland Public Library, Ms. Ethel L. Robinson, head
Denver Public Library, George De Luca, researcher
Des Moines Public Library, Ms. Shirley Shisler, Reference Department
Free Library of Philadelphia, William Felker, Information Department
Kansas City Public Library
Louisville Public Library, Mark Harris, Kentucky Division
Memphis/Shelby Co. Public Library, Daniel A. Yanchisin, Information uDivision
Milwaukee Public Library, Paul Woehrmann, Local History Division

National Climactic Center, Ashville, NC
National Research Council, Ms. Dorothy Kriete, research librarian
Omaha Public Library, Mrs. Patricia Bleick, Reference Department
Onondaga Historical Society, Mr. Richard N. Wright, director
Oswego City Library, Ms. Mary J. Nieder, director
Portland (Maine) Public Library, Local History Division
Seattle Public Library, Kim R. Turner, Newspaper Reference Division
Syracuse/Onondaga Co. Public Library, Gerald L. Parsons, Local History & Genealogy Division
Syracuse Government Research Bureau, Richard C. Spaulding, director
University of Alaska Library, Ms. Rochelle Sager, Reference Division
University of Wisconsin, Milwaukee, Professor A. Theodore Brown

Finally, because of the protracted character of this study, many of my benefactors listed above have no doubt advanced to other posts or retired, but my gratitude for their aid persists. And in the final preparation of my manuscript for publication I am indebted to Ms. Ruth Rosenberg Napersteck, who has succeeded Dr. Barnes as City Historian, for aid in the search for appropriate illustrations, and to Robert Easton, Managing Editor, for painstaking editorial help.

THE BENDED KNEE RESPONSE

The Rev. James MacSparren, in a sermon delivered at Narragansett in March 1741, pronounced:

> The recent suffocating snows [are] a warning of God's Vengeance on us for our Ingratitude to his Goodness and our Transgression of His law . . . Would we therefore be relieved of the Burden and Inconvenience of the winter . . . we must propitiate the God who alone . . . can invite us to sing and say, in the Language of Solomon's Song, "Lo the Winter is past . . . The Flowers appear on the Earth, the time of the singing of Birds is come, and the Voice of the Turtle is heard in the Land."

(Ludlum, *Early American Winters: 1604–1829*, vol. 1, pp. 244-45, quoting from a document in the Brown University Library.)

CHAPTER ONE

SNOWSTORMS IN PEDESTRIAN TOWNS

1620–1800

After successfully battling snow drifts, enduring traffic tie-ups, splashing through salty puddles, and bundling up against the frigid blasts of yet another harsh winter, many city-dwellers take delight in recalling earlier bouts with blustery storms. Journalists and meteorological broadcasters are quick to supply historical antecedents and reminiscent details of former sieges, thus gratifying the average American's sporting delight in records matched or surpassed—even records of hardships endured. Citizens throughout the Snow Country, bounded on the south roughly by the 39th parallel (and by the more generally acclaimed but not impenetrable Sun Belt), have never seen so many records broken as in the "standout winters" of 1976–77 and 1977–78, and by the "White Hurricane" that swept up the East Coast in March 1993. Many who have suffered through them, especially in hard-hit Buffalo and Boston enjoy a sense of heroism just for surviving their bruising fury.

But the real heroics, the real achievement was in the response, which broke all former records in both resident and civic efforts. Adequately to appraise that response, whether in these truly landmark winters or in other harsh seasons in recent years, it is necessary to look back not only to former record-setting winters, of which every community has had its share, but also to an earlier period when snowstorms were accepted as natural or God-given punishments to be complacently endured. Fortunately, sufficient information is available on the experiences of the "pedestrian" towns and emergent cities of Colonial America to reconstruct an ideal baseline from which to trace the slow development of diversified urban efforts to combat and survive recurrent snowstorms.

The first discovery in such an historical review is the close relation between the size and character of the community, the nature of the storm's disruption, and the quality of the response. Thus while the pedestrian towns of Colonial days suffered no traffic tie-ups, they could not escape the drifting snows and bitter cold of repeated harsh winters, and their response to these hardships supply historic antecedents for the increasingly complex snow-fighting efforts of the urban society that followed.

In view of the many other hardships they faced, it is comforting to learn, as David Ludlum tells us,[1] that a relatively mild winter favored the Pilgrims after their landing at Plymouth late in December 1620. Fortunately, too, another mild winter a decade later greeted the founders of Boston. William Wood, a member of the latter migration, observed this coincidence a few years later in his commentary, the *New England Prospect*. Writing to assure his countrymen at home, he described the climate of the New England coast as "agreeing well with out English bodies." An earlier group of settlers at Sagadahoc in Maine had, however, found the winter storms of 1607–08 more

[1] David M. Ludlum, *Early American Winters: 1604–1820* (Boston: American Meteorological Society, 1966). I am heavily indebted to David Ludlum who has painstakingly research the climatological records and diaries pertaining to colonial and early American winters. This and his succeeding volume (II:1821–1870) hereafter cited as *EAW*, comprise the major source for my first two chapters, in which my focus is on the social response rather than on the meteorological evidence and trends that chiefly interest him.

"vehement"; all survivors had sailed for home after the return of spring. Severe winter storms soon assailed both the Plymouth and Boston settlements but providentially (as most contemporaries saw it and many today would affirm) not until after they had erected some secure shelters and had become accustomed to the sudden changes in climate. William Wood, eager to present a favorable view, heard fewer men "in public assemblies sneeze or cough as ordinarily they do in old England."[2]

Some "Remarkable Providences"

A number of harsh winters soon dispelled this early complacency and recast the blustering storms as expressions of divine wrath. No doubt some colonials gave little credence to that interpretation, but accounts of those who survived exposure to a major snowstorm frequently attributed their good fortune to "God's mercy." Perhaps the tone of many reports was influenced by the clergymen who transcribed them, for many of the early journal-keepers were men of the cloth. But the Elect who trudged through deep drifts to reach a meeting-house on the sabbath during a heavy storm were not surprised to hear the pastor "pray for mitigation of the weather," as Judge Samuel Sewall confided to his diary during a protracted storm in February 1698.[3]

That was the third of the seventeenth century's "three landmark winters," as Ludlum describes them. A revealing sense of resident priorities emerges from an examination of Noah Webster's edition of *Winthrop's Journal*. Passing over the "unusually severe" winter of 1637–38, when the tiny Boston settlement was blanketed by a snow of eighteen inches, he identified 1641–42 as the first exceptionally

[2] Ludlum, *EAW*, vol. I, pp. 6–14.
[3] Ludlum, *EAW*, vol. I, p. 17, quoting from Samuel Sewall, *Diary of Samuel Sewall, 1674–1729* (Boston: Massachusetts Historical Society, 1878).

harsh winter, the worst, in the memory of friendly Indians, in forty years. More remarkable than the depth of the snow, which exceeded three feet in some outlying districts, was the severity of the cold. Not only was the bay around Boston frozen over, but reports arriving later told of a similar freeze as far south as Virginia where the rivers were solidly covered. In Boston, men drove horses and carts over ice that was six inches thick and sufficient to support loads of firewood hauled by oxen from the mainland. Several men lost their lives trying to bring boats into port at the onset of the storm, but a few weeks later, when a thaw re-opened the bay, "by the good providence of God" only one fell through the ice and was drowned. Clearly the chief concerns were to replenish the supply of firewood in town, and to assure the safety of the men who manned the barks and larger sailing vessels that normally maintained communications and carried supplies between the scattered settlements and to the homeland.[4]

Deep snows, such as those that fell "but lay not long" in Boston, already a town of twelve hundred inhabitants in December 1642, were not as serious for its closely clustered residents as the widespread storms that sometimes blanketed inland towns for several weeks and resulted in the loss of cattle trapped in barren shelters. Men suffered frozen feet and hands while performing essential chores in town as well as country, but heart failures from snow shoveling were unrecorded.

Governor Winthrop sometimes noted the suspension of the courts, because of deep snows, as at Boston in February 1645, but with other weather watchers, he displayed greater concern when a winter gale coated ships riding in the harbor with ice, breaking their moorings and driving them ashore. John Hull's account of a storm that struck in January 1667, when it snapped the cables of eight ships in the Boston harbor, four preparing to sail for England,

[1] Ludlum, *EAW*, vol. I, pp. 15, 18–21, 33.

highlights the concern many felt for the preservation of the port's vital ties with the mother country. Similarly in Virginia, the most damaging weather condition, as reported to Winthrop, was suffered in 1645–1646, when "the ships were frozen six weeks" in the bay.[5]

The diary of Judge Samuel Sewall of Boston supplied revealing details of several harsh seasons. He briefly described the winter of 1680–81 as "a very severe winter for snow and a constant continuance of cold weather." The Rev. Increase Mather, who included it among his "Remarkable Providences," labeled it "The coldest winter that has been known these 40 years." Neither, however, observed specific reactions to the storms in Boston or elsewhere, possibly because the monthly postal services opened to New York, in January 1673, had long since succumbed to earlier storms.[6]

Luckily, almost two decades later, "the terriblest winter" of 1697–98 received more adequate coverage. From the clerk of Sudbury town, who gave it that title, to the Earl of Bellmount, officially in command at New York, its severity was widely attested. The Rev. John Pike, the intrepid weather watcher in Dover, New Hampshire, counted thirty-one snowy days that winter. A severe storm on December 30 prompted Cotton Mather to preach on the familiar text, "Refuge from the storms of the Wrath of God." Judge Sewall supplied fuller details. His entry on February 2 reported "very thin assemblies this Sabbath and last; and great coughing; very few women there; Mr. Willard prayed for mitigation of the weather...." On March 3 he recorded a visit to Charlestown where he found that after the Rev. Wigglesworth had preached on February 2 on the text, "Who can stand before His cold," the church had suspended services for three weeks. Indeed, judging

---

[5] Ludlum, *EAW*, vol. I, pp. 19–21, 23, 33.

[6] Samuel Sewall, *Diary*, pp. 466–75; Carl Bridenbaugh, *Cities in the Wilderness* (New York: The Ronald Press Company, [c. 1938]), p. 53.

from the available reports, a suspension of church services had become the accepted gauge of a severe storm in scattered interior towns if not in the leading ports. In nearby Cambridge, the First Church apparently kept open that winter, but its clerk recorded snow "three feet and a half deep on a level," which had prompted the church to schedule a public fast on March 27 to petition for relief.[7]

When the snows melted, the colonial ports resumed their growth and as a result developed several new wintertime concerns. A few private carriages and the first hackney coaches made their appearance in Boston and New York in the late 1660s as these towns acquired populations of 2500 or more; when snow clogged the streets a few sleighs took their place, but most men traveled by foot or horseback in town or on overland journeys. When the inter-colonial postal service resumed in the 1690s, the post riders linked Boston, New York, and rapidly growing Philadelphia, and extended the weekly overland routes north to Portsmouth, south to New Castle, PA, and west to Lancaster before the close of the century. The appearance of the first weekly *News Letter* at Boston in 1704 demonstrated its urban leadership at that date; it also served to institutionalize the demand for regular postal services and supplied a public outlet for the observations of scattered weather watchers who now increased in number.[8]

Two severe winters, 1704–05 and 1705–06, brought the first published complaints at Philadelphia and New York over delays in the arrival and dispatch of the post. Deep snows at Philadelphia, spreading northeastward to New York and New England, halted the postal service for six weeks early in 1705, according to Isaac Norris of Philadelphia. Successive falls, totaling three feet in some reports and drifting to impassible depths in many places, blocked

---

[7] Ludlum, *EAW*, vol. 1, pp. 16–17.

[8] Bridenbaugh, *Cities in the Wilderness*, pp. 22, 198, 204–05; Wesley E. Rich, *The History of the U.S. Post Office to 1829* (Cambridge: Harvard University Press, 1924), pp. 12–24, 29.

all travel over the recently opened roads. Fortunately the severe cold coated the Delaware and the Hudson rivers as well as lesser streams with a sufficient covering of ice to enable lumbermen to bring sled-loads of firewood to the shivering townsmen and to encourage other travel along the wind-swept rivers.[9]

The *Boston News Letter*, cut off from overland despatches that season, was able to report more fully on a similar series of storms that blanketed New York and Boston a year later. Periods of thaw in the winter of 1705–06 enabled the post riders to maintain their services, although somewhat irregularly. Concern for the fate of sailors manning the coastal ships continued but failed to ward off a catastrophe in December 1705 when the privateer, *Castle Del Key*, was blown aground by a gale in New York Bay, and 132 of the 145 men aboard perished in the icy storm.[10]

When the storms subsided, leaving a beautiful white blanket adorning the landscape, affluent townsmen in increasing numbers hitched up their sleighs for a brisk ride over the snow-covered streets. A British visitor to New York in 1704 passed "50 or 60 sleys" on one cold ride that winter. Philadelphia soon required all sleighs to have bells attached to the shafts or harness in order the warn the more numerous pedestrians of their approach. In Boston when the snow melted, generally in March, the selectmen launched a spring clean-up. Householders quickly swept the winter's accumulation of garbage in front of their properties into piles to be picked up by the cartmen when they made their intermittent rounds. A late storm sometimes brought complaints from the scavengers against householders who failed to "keep clear a good and sufficient path" to expedite the removal of their wastes. Such

---

[9] Ludlum, *EAW*, vol. 1, p. 41; *Annals of Philadelphia & Pennsylvania in the Olden Times* vol. 2 (Philadelphia: E. S. Stuart, 1898), p. 349.

[10] Ludlum, *EAW*, vol. 1, pp. 42, 241–42; *Boston News Letter*, Dec. 31, 1705 and Jan. 7, 1706.

problems were, however, buried and suspended when the big snows arrived.[11]

Boston and other New England towns bore the brunt of the "Great Snow" of March 1717. A succession of storms made deposits that totaled between 32 and 42 inches in the city, according to varied reports, and piled up drifts of 20 and even 25 feet in some rural districts. Again the posts were delayed for two weeks, and when a post rider from New York finally broke through, he arrived on snowshoes, reporting the interior impassible for horses. The *News Letter* maintained its weekly schedule with limited editions despite the snow, which "lies in some parts of the streets about six feet high." Drifts against the building in some cases covered second-story windows. After commenting with awe on the first storm, Cotton Mather reported of a second" "Another snow came on which almost buried ye Memory of ye former, with a Storm so famous that Heaven laid an Interdict on ye Religious Assemblies throughout ye Country." His own church suspended services for two Sundays in a row, and Mather proposed " a day of humiliations and supplication" and, in the next entry, pledged a "suitable expression of charity."[12]

Other heavy snows followed, but it was the winter of 1740–41 that achieved landmark status. Marked by frozen harbors from Maine to Virginia and by snows that exceeded three feet at several places in New England, the "Hard Winter," as it was to be remembered, dropped the temperature at Charleston, South Carolina, that December to the lowest monthly average since thermometer records there had started. That in fact was not a long span, since thermometer readings there had begun only four years before, but this was the first winter that brought thermometer readings into the weather reports from

[11] Bridenbaugh, *Cities in the Wilderness*, pp. 274, 322.
[12] Ludlum, *EAW*, vol. 1, pp. 42–46, 242–44, quoting the *Boston News Letter* and the diaries of Judge Sewall and Cotton Mather among other sources.

scattered locations—Philadelphia, Cambridge, and New Haven as well as Charleston—which made it a landmark winter in a meteorological sense as well.[13]

Urban Snow Hazards Emerge

That Hard Winter had a wide and sometimes devastating impact. It exacted a toll of "50 sail" in Massachusetts Bay, halted shipments from Boston for thirty days and on the Delaware for almost three months. It dropped five inches of snow on Savannah where, however, it melted within a few hours. It brought new weather-watchers onto the scene, as well as weekly journals at Philadelphia and New York, and elicited the first comments on urban storm conditions. "Our streets are confused with heaps of snow, so that the lovers of sledding can scarcely use them without danger," observed the *New York Weekly Journal*. When the storm abated, sleigh riding on ice-covered rivers and along the north shore of the Sound and the south shore of Chesapeake Bay received notice in varied reports.

Only the more affluent enjoyed such pastimes, while most poor house holders suffered severely from the protracted cold. The wider growth of Boston and the other burgeoning ports had absorbed the forest lots that had formerly provided firewood, depleting the supplies and forcing prices up to forty shillings a cord by March in 1741. Many, unable to pay such prices, appealed for relief, and the Selectmen of Boston voted to spend £700 to stock a wood yard where poor residents could obtain a small quantity to re-light their fires. New York, faced by similar hardships, sponsored a collection that raised £500 to supply wood and coal to the needy.[14] Thus in the three leading cities, each now boasting over 10,000 inhabitants, the wintertime

[13] Ludlum, *EAW*, vol. 1, pp. 48–50.
[14] Ludlum, *EAW*, vol. 1, pp. 49–51; Bridenbaugh, *Cities in the Wilderness*, pp. 213, 322.

concern for an adequate supply of fire wood acquired a measure of public backing.

The traditional attitude that accepted the widely-spaced severe winters as acts of God was still frequently voiced by small-town clergymen. The Rev. James MacSparren, in a sermon delivered at Narragansett near Providence, in March 1741, saw the recent "suffocating snows" as a warning of God's "Vengeance on us for our Ingratitude to his Goodness and our Transgression of his Law.... Would we therefore be relieved of the Burden and Inconvenience of the Winter... we must propitiate the God who alone... can invite us to sing and say, in the Language of Solomon's Song, 'Lo the Winter is past.... The Flowers appear on the Earth, the time of the singing of Birds is come, and the Voice of the Turtle is heard in the Land."[15] The clerk who kept the records for First Church in Cambridge during the next winter of heavy snows, 1747–48, must have been a man of good standing in Heaven. After briefly recording the 30 snows that had covered the ground there to a depth of 4 or 5 feet from December well into March, he was able to conclude brightly "on the twelfth day of April, I had my garden sowed and planted with onions, carrots, parsnips, peas and beans."[16]

Few other weather watchers in that winter of record-breaking snowstorms achieved such serenity. Dr. Edward Holyoke's diary recorded "snow on a level of 30 inches" in Salem on March 4 and added "no traveling about the country except on rackets." Joshua Hempstead of New London, after tabulating twenty snowy days in January and February, recorded a "violent storm" on March 2, which cut the attendance at meeting the next day to "a Thin Congregation, no women." To break a path before his door he hitched "two horses, backward and forward,

---

[15] Ludlum, *EAW*, vol. 1, pp. 244–45, quoting from a copy of the sermon printed in 1741 and on deposit in the Brown University Library.

[16] Ludlum, *EAW*, vol. 1, p. 55, First Church Records, *Genealogical Magazine*, vol. 1, p. 361.

before his sled" in order to buck the drifts. Men on essential errands that March often dismounted and led their horses through the towering snowbanks.[17]

A succession of relatively mild winters enabled the emergent cities to organize their services more effectively. As each fall approached, a fleet of small boats carried firewood from Maine to Boston, from Connecticut and Long Island to New York, and from the upper Delaware region to Philadelphia. When snow and ice blocked the ports, drivers from nearby hamlets brought sleds loaded with firewood into town to take advantage of the higher prices. During a frigid spell, in January 1760, "upwards of 1000 sleds" reached Boston in one day. The volume of traffic in the streets was mounting as the private and hackney coaches, carriages, and chaises increased, exceeding eighty in New York, then third in size among the colonial cities, by the mid-sixties. A heavy snow frequently prompted the owners of some of these vehicles to replace the wheels with long runners, while other townsmen in increasing number kept sleighs conveniently at hand in their stables.[18]

December 1764 brought heavy snows to several east-coast cities. In Philadelphia a sufficient number of sleighs turned out to maintain a lively traffic well in January despite a fall of two feet in suburban districts. An entry in the Note Book of James Watts, later secretary of the New York Historical Society, noted "snow and sleighing in abundance, more than has been know for this twenty years."[19] Sleighing waned as the storms became more severe, increasing the depths to "near three feet" at Hartford and several other places. Ice again blocked the Delaware River and at New York clogged the Narrows, obstructing use of the harbor. Floating cakes of ice in Boston's harbor, driven by waves and tides against the wharves, wrecked many

---

[17] Ludlum, *EAW*, vol. 1, pp. 55–60.
[18] Bridenbaugh, *Cities in the Wilderness*, pp. 26–27, 34–35, 146, 341, 371.
[19] Ludlum, *EAW*, vol. 1, pp. 60–63.

docks and carried off ship timbers, firewood, and other goods, causing great damage to the port as well as to ships at anchor in the bay. In New York City, where many inhabitants kept cows in their barns, engaging herders to drive them daily to and from suitable pastures, the deep snows and severe cold of January and February that year interrupted the regimen and presented new provisioning problems to their owners. A March storm struck Portsmouth, New Hampshire, battering its ships and warehouses and inflicting losses "supposed to amount to some Thousands," according to the *Portsmouth Mercury*.[20]

Another snowy March spread a heavy blanket over much of the Northeast in 1772, but the harshest blows that winter fell on the South. A great snowstorm, raging over Virginia and parts of North Carolina in January, deposited as much as three feet in some districts according to diary entries by both Washington and Jefferson. The meeting of the General Assembly at Williamsburg had to be postponed. The storm, which exceeded anything known to that region in a hundred years, dumped snow as far south as Savannah, though it soon melted.[21]

Snow Hazards of the Revolution

Residents of the colonial ports and the scattered inland towns had learned by the mid-seventies how to endure if not to cope with the heavy snows and harsh temperatures of American winters. The experience would prove a great asset as they challenged the might of the British armies during the Revolution. That contest was not primarily an urban one, but the British, who tended to base their strategy on holding the cities, failed at crucial points to prepare

[20] Ludlum, *EAW*, vol. 1, pp. 60–63; John F. Watson, *Annals and occurrences of New York city and state, in the olden time* (Philadelphia: H. F. Anners, 1846), p. 199.
[21] Ludlum, *EAW*, vol. 1, pp. 63–64, 144–146.

for wintertime moves by their opponents. The Americans were able on several occasions to take effective advantage both of snowstorms and of thaws and in the process increased their knowledge of, and capacity to cope with, winter snows.

The most crucial use of winter weather occurred during the siege of Boston. General Washington, with ill-equipped and untrained troops, was unable to assault the entrenched and experienced British forces without increased armament. Fortunately a supply of cannon had recently been captured at Ticonderoga, some 260 miles inland. The mid-winter transport of those cannon over forest covered hills and valleys by Colonel Henry Knox and his men, with the assistance of some eighty yoke of oxen, was only accomplished with the aid of the below-freezing temperatures and substantial snow coverage that the moderately severe winter of 1775–76 supplied. General Howe, resting comfortably in Boston, awoke too late to his danger and had to withdraw.[22]

Washington's familiarity with the changing character of the winter storms proved advantageous on other occasions. His dramatic crossing of the Delaware on Christmas Eve in 1776, and his escape a week later from a trap by Cornwallis, both demonstrated an ability to react quickly to shifting weather conditions. These actions had little relation to the urban response to snowstorms except perhaps to demonstrate the importance even to rural encampments of maintaining open routes of supply and communications under varied weather conditions.[23]

This was also a major concern for the British in New York and for the Continental forces in the varied ports they held, especially during the hard winter of 1779–80. The deep freeze and heavy snows of that third "Landmark Winter" effectively closed the New York harbor, and all

---

[22] Ludlum, *EAW*, vol. 1, pp. 96–98.
[23] Ludlum, *EAW*, vol. 1, pp. 98–99.

other ports north of Annapolis as well. It considerably immobilized both armies but cost the British more dearly in prestige and morale. Unable to bring in supplies by sea, the British commandeered local stores of grain and other foods and chopped down most of Manhattan's trees to keep the troops warm. When the East River froze over, permitting a crossing with ice sleds, they despoiled parts of Long Island as well. These actions and the accompanying hardships weakened the morale of some of the Loyalist supporters who had flocked to the city to take the places of patriots who had withdrawn when the British arrived. The Continental forces, by contrast, suffered more deprivations but maintained communications along the ice-covered rivers. In New England, Boston, Worcester, and Hartford managed to keep one overland road open by repeated efforts to break through the shifting drifts with the aid of double teams of horses led sometimes by men on snowshoes. New Haven replenished its supply of firewood that January by mustering fifty men with horses and sleds to break a path six miles to an outlying wood yard.[24]

New England towns enjoyed a more regular postal service that winter than many in the South where the depth of snow presented an unaccustomed obstacle. But the northern journals faced a new hazard as the freeze halted water-powered paper mills and reduced the supply of paper, compelling the *Boston Gazette* among other weeklies to limit some editions to a single page in January 1780. Freezing weather halted paper mills in the South as well, and some of the more affluent inhabitants there found a new delight in sleighing over the frozen surface of Chesapeake Bay, venturing in a few instances as far south as Norfolk harbor.[25]

The return of peace brought a resurgence to the cities,

---

[24] O.T. Bark, *New York During the War for Independence*, (New York: Columbia University Press, 1931), pp. 108–17; Ludlum, *EAW*, vol. 1, pp. 85, 111–16, 122, 128.

[25] Ludlum, *EAW*, vol. 1, pp. 116, 148–49.

most of which had suffered a decline due to their occupation of because of fear of attack. Six rival ports joined the three leaders and three inland towns, most of them in the snow country; each boasted two or more weeklies, over 10,000 inhabitants, and a mounting congestion of horse-drawn vehicles that now outnumbered the boats in most harbors. A new contingent of weather-watchers had also arrived; equipped with thermometers, barometers, and clocks they were more concerned with collecting specific readings and comparing them with past records and with neighboring reports than with speculating on a Heavenly message.[26]

Successive snowy winters in the mid-eighties presented an early test of citizens' new mettle. The record-breaking length of the winter of 1783–84, the late but deep snows of 1785, and the triple storms of 1786, each spread heavy blankets over New England and dumped drifts south into Pennsylvania and Virginia. Scattered weather-watchers reported sub-zero temperatures in several towns, notably at Hartford where, in February 1784, according to Noah Webster, they fluctuated from −12 to −20 for eight mornings in a row. A report from Baltimore, where the harbor was closed for two months, told of a successful effort in March to re-open it with "much labor in cutting a passage." In the heavy snows of the next winter, wagoners converted to using sleds, and by driving them in tandem, broke open the roads; weather diarists reported "good sleighing" more frequently than in earlier years, indicating a widespread accommodation to the winter seasons. Stage coaches mounted on runners maintained service between New England towns throughout the winter months.[27]

The last decade of the 18th century brought two memorable winters—a cold January in 1792 and a long

[26] Ludlum, *EAW*, vol. 1, pp. 76–77.
[27] Noah Webster, "Notices of Extraordinary Seasons of Cold," *American Journal of Science* (July 1835): pp. 183–86; Ludlum, *EAW*, vol. 1, pp. 64–73, 152.

winter in 1798–99. Each blanketed a wide area in the Northeast, dropping heavy snows on many towns, but the increased use of horse sleds and sleighs enabled the inhabitants to keep most streets open in the cities and to reopen main roads through the villages. A ten-inch snow frequently brought comments on the "find sleighing" that resulted. The January 1792 storm extended as far west as Cincinnati where the pioneer settlers, overlooking a river blocked by ice for thirty days, had to wade through two feet of snow in search of supplies in the back country. In 1798, storms caught numerous vessels off the New England coast off guard and cast seven wrecks on the shores of Cape Cod with the loss of many sailors; but the cities, no longer so dependent on the coastal trade, suffered less. Long experience had prompted many householders to lay in a supply of firewood, and numerous wood dealers stood ready in every town to serve customers with the aid of horse sleds or carts as conditions dictated. Hackney coaches, mounted on runners in snowy weather carried residents as well as visitors about the larger towns. The era of the "pedestrian cities" was passing in the Northeast and new wintertime concerns were emerging.[28]

---

[28] Ludlum, *EAW*, vol. 1, pp. 73–76, 154–62, 225–26.

FOUR ENGINES PUSHING AN IMPROVED PLOW
(*Harpers*, 1860)

The stoical acceptance of heavy snow, packing it down for smooth travel on runners, was first challenged in the 1830s by the new inter-city railroads with horse-drawn V plows, which spurred the development of giant steam-driven plows.

CHAPTER TWO

SLEIGHBELLS AND STEAM WHISTLES IN THE SNOW

1800–1870

If some eastern cities had outgrown their pedestrian, snowshoe days by 1800, many new towns in the interior would repeat those earlier experiences for another decade or so. Their ordeals would expand America's knowledge of harsh winter storms, many of which now were seen to surge out of the West, rivaling, indeed overshadowing, those that still roared up the coast. As the number of thriving towns increased, a host of new weather watchers emerged, and after 1821 many would send their reports to the Surgeon General's Office in Washington, and later to the Smithsonian Institution where a tentative effort to analyze national weather trends began. A multitude of weekly and finally daily newspapers sprang up, each eager to publicize the more dramatic local and distant records. The stage drivers, who had displaced the post riders, increasingly gave way in the forties to the conductors of steam trains as the carriers of postal bags, and they in turn lost out even more quickly in the fifties to telegraph agents in the dispatch of weather records and other urgent communications.

These were but a few of the technological advances that created new urban wintertime concerns. Chugging engines, often bucking heavy drifts in tandem, tooted and tolled through the blinding snowstorms to warn pedestrians and teamsters away from grade crossings; steamboats caught in surging blizzards plaintively whistled of their plight at lake and river ports, hoping often in vain for a signal light; and as hackneyes and urban omnibuses gave way to horse-car lines in several cities, the seasonal switch to sleighs was complicated by each company's efforts to clear its tracks.

Old Fashioned Winters

Yet, despite their continued ferocity, the snowstorms of the first seven decades of the 19th century appear, at least in the retrospective light of urban reminiscences, to have acquired a more benign quality than those of the colonial period. Perhaps the reason was that inhabitants of new towns and old cities alike had acquired greater self-assurance. Heavy snowfalls brought advantages as well as hazards. The opportunity to move bulky loads of produce and fragile imports on great sleds over forest roads to and from distant ports facilitated the growth of inland towns. The delights of gliding over the snow-packed streets and rural roads prompted residents of towns and cities to defer distant journeys until the snows arrived; after each storm subsided, many joined their neighbors in hitching horses to heavy sleds to break open the roads for merry and jingly sleigh rides about town. Snow-bound editors welcomed reminiscent accounts comparing the present storm with local and distant predecessors, and took apparent delight when an old settler could exclaim, as Edwin Scrantom did over a succession of storms at Rochester in 1839–40, "Vermont is beat all hollow." The city's 20,000 inhabitants were virtually immobilized. The only activity Scrantom mentioned was that of burying "Brother Timothy Haskell,... a last service performed at his request by his close friends but with great difficulty."[1]

The snowstorms, however, had not lost their punch. Even in the relatively mild winter of 1801–02 a storm that hit New York City on February 21 acquired blizzard proportions as it spread over New England. Heavy snows fell on New Haven and Cambridge, and the wind piled up great drifts inland that reached depths of eight and even twelve feet at Exeter. Rutland reported an accumulation of four feet on a level, and Worcester forty inches, enough to provide a base for good sleighing until the second week in March. A somber report from Cape Cod announced the loss of nine of fourteen men on a ship driven ashore during the storm. The winter of 1804–05 packed the harshest blows. A snow hurricane on October 9, 1804, dropped twenty inches in Essex County, immobilizing New London, but failed to reach Boston or Providence. A great storm in January was more widespread. It gave New York City perhaps its heaviest single snowfall, over two feet in depth, and deposited forty-three inches on New Haven. Noah Webster dubbed it the severest winter since 1780. In New York, now the country's largest city with over 70,000 inhabitants, the streets were impassable; drifts reaching five feet clogged all traffic. Only one stage made the short distance between Cambridge and Boston during the first week, and the turnpike north to Lynn and Salem was opened only "with great difficulty." Yet a few days later reports of "good sleighing" arrived from many places.[2]

These were Eastcoast storms, as David Ludlum has reconstructed them, and December 1811 brought a similar onslaught, which blanketed the Boston area with from

---

[1] Blake McKelvey, "Snowstorms and Snow Fighting: The Rochester Experience," *Rochester History* XXVII (Jan. 1965): p. 3.

[2] Ludlum, *EAW*, vol. 1, pp. 165–66, 169–74. Not until Dec. 26–27, 1947, would another storm (26.4 inches) rival and possible exceed the New York snowstorm of Jan. 26–27, 1805.

twelve to sixteen inches of snow and spread inland to provide good sleighing in January throughout most of New England. But reports from a scattering of weather watchers in the West identified a raging storm on April 1, 1807, as springing out of the lower Ohio valley. As it advanced up the valley and continued across Pennsylvania, it dropped fifty-two inches on Scranton and fifty-four at Utica. New York received only six inches but suffered severely as the gale battered ships in its port. In Vermont, the *Randolph Weekly Wanderer* blamed its failure to publish the regular issue of May 6 on snows that "made about six feet on a level."[3]

In the past, masses of cold air from Canada had helped to transform coastal storms from the South into raging blizzards, but few cold waves rivaled that of 1816. Its frigid sway effectively warded off southern storms. Relatively light snows off the lakes and from the west dropped scattered deposits throughout the Northeast during January 1816, "the year without a summer." An arctic wave from Canada, dipping down through northern New York and New England, held the entire region in a chill that stunted crops and created an unending demand for firewood as far south as Philadelphia; it also brought coal into wider use in several cities. A succession of snowstorms in the northernmost towns in June proved, however, to be the year's most memorable events,[4] adding reminiscent interest to their histories.

Snowstorms continued their periodic visits throughout the Northeast, but the response in towns and cities was changing. Local editors hailed an "old fashioned winter" with a degree of zest. "The people are out generally breaking roads," became a common report; even the Rev. Thomas Robbins of East Windsor near Hartford noted in his diary on February 10, 1820, that he had "worked

---

[3] Ludlum *EAW*, vol. 1, pp. 175–78, quoting the *Weekly Wanderer*, May 13, 1807.

[4] Ludlum *EAW*, vol. 1, pp. 182–86, 190–93.

shoveling snow two hours." At Boston a decade later, when a January storm left two feet of snow in its wake, the *Boston Gazette* reported confidently: "Several sleds filled with men and boys, attached to each of which were six horses, were driven through the streets of the city yesterday afternoon to break and level the drifts. Our truckmen understand this business, they practice the *leveling system*, on all great occasions, with more success than any other class of citizens." Not only were many urban residents looking forward to sleighing; some were actually enjoying the effort to make it possible.[5]

Many were enjoying the winter seasons in other ways as well. With increasing populations, a greater number of horses and sleds, and more ample fuel supplies to keep stoves and fireplaces blazing, towns and cities acquired a merry atmosphere in winter seasons. When the temperature dropped sufficiently to coat the Hudson River with a solid covering, as it did in January 1821, enterprising New York merchants opened "refreshment taverns" on the ice to serve those who ventured out for sport or on a trip to New Jersey on business or pleasure.[6]

The "Great Snowstorms" of January 1831, however, presented difficult problems to many towns. A fall of twenty-two inches at Pittsburgh and of thirty inches at Gettysburg not only clogged their streets but effectively isolated them from eastern mails. The stages were in fact halted at Lancaster where a concerted effort with fifty horses and many men failed to open the highway blocked by drifts ten and twelve feet deep. Far to the east, New Bedford was isolated by six-and eight-foot drifts. Even Baltimore, on the southern edge of the storm received eighteen inches, but its recently chartered and partially constructed steam railroad, the Baltimore & Ohio, success-

[5] Ludlum, *EAW*, vol. 1, p. 197, quoting the weather diary of Rev. Thomas Robbins; Ludlum, *EAW*, vol. 2, p. 15, quoting the *Boston Gazette*, Jan. 17, 1831.

[6] Ludlum, *EAW*, vol. 2, pp. 3–11.

fully cleared the tracks for its trains by sending a team of horses ahead with a "contrivance" to push the snow aside. That historic introduction of a snow-plow and of a steam train to supplant the inter-city stages marked the first glimmer of a new era in transportation and in snow-fighting as well.[7]

It was, however, only a glimmer, and farther north the heavier snows and harsher gales soon extinguished it. New York and Philadelphia had turned to canals for most of their commerce, and both were more concerned over the dates when the Hudson and Delaware Rivers, and now the Erie Canal were closed to traffic. When a cold wave late in 1831 closed the Erie Canal on December 1, it was widely remarked as the earliest closing in its history. Five years later, when both the canal and the Hudson River were closed to traffic on November 30, the event was hailed as a new, if unwelcome, record. Few were surprised in February to hear that the newly-opened Camden & Amboy Railroad in New Jersey, connecting New York with Philadelphia, was snow-bound for five days. When the train, drawn part way by horses, finally reached South Amboy, the passengers found the steamboat scheduled to carry them into the city frozen in its slip. In Massachusetts, the Boston and Worcester boasted of maintaining its forty-mile service on all but six days of that winter. But if rail service was interrupted, and steamboat and ferry service was suspended for several weeks in the frigid winter of 1835–36, residents of Boston and New York enjoyed the excitement of venturing out as pedestrians to walk on ice-covered rivers and bays; refreshment stands again served hardy customers in the middle of the Hudson River.[8]

---

[7] Ludlum, *EAW*, vol. 2, pp. 99–100, 224; *American Journal of Science* 20 (Jul. 1831): p. 166.

[8] Ludlum, *EAW*, vol. 2, pp. 16–27. David Ludlum lists fourteen winters from 1696–97 to 1874–75 when pedestrians crossing on the Hudson were reported. *See also* "New York Weather Highlights," *Weatherwise* (Apr. 1961): p. 63.

Sleighbells and Steam Whistles in the Snow

1836 was, after all, a "standout winter," particularly in upstate New York. At Rochester, the "Big Snow," as Professor Chester Dewey described its February blizzard, left a deposit of three feet; at Utica, four feet. News of the collapse of several buildings in the latter city prompted its mayor to call an emergency meeting of the councilmen and leading merchants to plan remedial efforts. The council ordered householders to clear their roofs and awnings and approved modest funds to help open the streets—the first such municipal action yet discovered. At nearby Syracuse, the "breaking out of the streets was delayed," but at Rochester enterprising cartmen, assisted by a partial thaw, soon had the drifts broken and the streets hard-packed with an eighteen-inch base ready for good sleighing. Residents of New York City enjoyed sleighing for three months that winter; although most of its east-west streets remained impassible for many days in January, the avenues were open and crowded, with sleigh bells contributing a note of merriment. Snowball fights, as reported by the *Herald*, broke out at nine o'clock closing time on successive evenings between the clerks and the hackmen on Broadway.[9]

The newly chartered cities of Brooklyn, Utica, Buffalo and Rochester, had each adopted ordinances requiring householders to build sidewalks and to keep them free of obstructions, but like the older cities in the East, none had yet made concerted efforts to clear snow and ice from these uneven paths. The inhabitants, particularly on the major streets, had a joint interest in opening the roads for sleds bringing in produce and for stages maintaining distant connections. After a heavy storm, which dropped an unprecedented total of four feet on Rochester in January 1839, residents along the highway leading to the docks dug a tunnel through the drifts in one blustery section and attracted other residents out to see it, prompting many to

---

[9] Ludlum, *EAW*, vol. 2, pp. 29–32, 215–16, quoting the *New York Herald*, Jan. 11, 1836; McKelvey, "Snowstorms and Snow Fighting", pp. 2–3.

25

open lanes in their own streets. The mails were suspended, but husbands and sons of the Female Charitable Society ladies carried packages of food and warm clothing to the needy.[10]

As the number of short inter-city railroads increased in the 1840s, the reports of snow blockades on "steam roads" mounted. The New Haven & Hartford, halted by snowdrifts in 1839, was stalled again in March 1843, as was the Auburn & Syracuse, in its case by twenty-five-foot drifts. That storm, which extended as far south as Philadelphia, clogging its streets and halting most traffic, failed however to cancel a parade by the Hibernian Greens, who waded and stumbled down Chestnut Street with a determination, if not a formation, that won commendation. Neither railroads nor stages could be expected to make much headway across a landscape described by one observer as "a vast ocean of snow" reaching from Albany to Buffalo. A second great storm ten days later spread the wintry tide eastward over most of New England. That was the tail end of the long "Hard Winter" in Wisconsin, when the snow blanket reached a depth of thirty inches at Milwaukee, burying rail fences and prompting boisterous sleigh rides across lots on the city's outskirts. It would be recalled forty years later as an era when "the loss of a week or two of time was of little consequence" and when "the tales of a wayside inn compensated for a snow blockage."[11]

Urban Snow Hazards Emerge

But if some citizens continued to accept the rigors of winter storms with a degree of zest, others, particularly in rapidly growing cities, were experiencing increased hardships. Frigid blasts delayed the delivery of much-needed

---

[10] *Rochester Daily Democrat*, Jan. 28 and 29, 1839.
[11] Ludlum, *EAW*, vol. 2, pp. 35–40; *Milwaukee Sentinel*, Mar. 13, 1881.

fuel and food supplies hazardous to those in short supply, aggravating the plight of many poor residents and revealing sad deficiencies in welfare services. Thus in the hard winter of 1805, Mayor DeWitt Clinton of New York, appalled by the plight of "ten thousand souls" whose supplies were approaching exhaustion, had petitioned and secured aid from the state legislature to open emergency wood yards and soup kitchens. Private charity societies redoubled their efforts, and when the successive storms halted street and other construction projects, throwing many day laborers out of work, a citizens' mass meeting organized a Samaritan Society to coordinate the collection and distribution of donations by volunteers in each ward. Although that society dissolved when spring arrived, other relief agencies appeared in subsequent harsh winters there and in other cities, and in 1817 the Society for the Prevention of Pauperism, which distributed 103,000 quarts of soup during a protracted storm in February, determined to be better prepared for future emergencies, and established a fuel fund that provided ample wood supplies in the harsh winter of 1820–21.[12]

All cities faced increased fire hazards in severe winters, but mounting congestion was now aggravating the problem. Fires from overheated stoves and chimneys were a regular occurrence, as was the inadequacy of the hand-pump fire engines with which volunteer companies endeavored to check, if not to extinguish blazes. A Philadelphia company blamed its failure to halt a fire at the French consul's house in January 1780 on the rapid freezing of the water in its tank before the pumpers could squirt it on the fire. This common hazard persisted and must have plagued many companies in frigid winters. On December 16, 1835, when a blustery storm hit New York City, fanning the flames of a fire in the commercial section, the fire

[12] Raymond A. Mohl, *Poverty in New York: 1783–1825* (New York: Oxford University Press, 1971), pp. 17, 20, 108–10, 251–52.

companies, equipped with the latest multiple hand pumps, again saw the water congeal before it left the nozzles of their hoses. It was this emergency that possibly prompted a foundryman on Wall Street to patent and build a steam fire engine in 1840, but the cost seemed prohibitive and despite his claim that it could supply an effective stream of water even in freezing weather, the demand was minimal.[13]

Ice often posed greater wintertime problems to cities than heavy snows. This had long been the case for the northern ports, which were sometimes ice-bound for weeks at a time. Boston merchants determined in January 1844 to cut a channel through the ice-blocked bay to enable ships locked in the harbor to commence their journeys. After a first attempt by 140 men to open a channel failed, a second, directed by the city's ice companies and supported by 500 men with ice plows and horses, succeeded and cleared the way for the departure of the steamer *Britannia*, which carried, in addition to fifty-four passengers, packets of mail said to contain 30,000 letters. A *Boston Post* reporter joined the multitude of citizens who ventured on the ice to cheer it on its way.[14]

As the relatively moderate snowstorms of the late 1840s gave way to the heavier onslaughts of the fifties, cities frequently reported a partial or total blockade of their railroads. Milwaukee was effectively snow-bound by a blustering storm in mid-January 1855. Chicago was for the first time completely cut off for almost a week by the same storm. With three engines in line, the Galena Railroad endeavored to break through the drifts but made only eight miles before returning for the night. Chicago's streets were "well blocked up,...awnings and signs torn from their fastenings,...and roofs caved in" by the weight of three feet

[13] Watson, *Annals of Philadelphia*, vol. 2, p. 357; Bart H. Vanderween ed., *Fire Fighting Vehicles: 1840–1950* (New York, 1972), pp. 8–9; Ludlum, *EAW*, vol. 2, pp. 21–24.

[14] Ludlum, *EAW*, vol. 2, pp. 218–19, quoting the *Boston Post*, Jan. 31, Feb. 5, 1844.

of snow. "Nobody braved the fury of the gale who could stay indoors," reported the *Chicago Daily Press*. Springfield was similarly snow-bound, with trains "marooned in the prairies." Its telegraph lines were down, its church services suspended, but one resident, Abe Lincoln, ventured forth to buy a pair of "overshoes, a small shawl, and for his wife two small combs and cotton flannel."[15]

Rochester, the birthplace of Western Union and already a promotional center for the spreading network of telegraph lines, had discovered three years before that freezing gales posed a great hazard to the fragile wires then in use. The frigid temperatures and strong winds of the Christmas season in 1852 not only downed lines and created consternation among local investors, but also inconvenienced editors who were cut off from their newly-extended news sources. The task of repairing and reopening the lines was hastily assumed by the telegraph and railroad companies whose prosperity, like that of earlier stage lines, depended on an early resumption of service.

But while heavy snow or a protracted deep freeze could immobilize shipping in lake and river ports, such as Buffalo, Cincinnati, Philadelphia, and Boston, it could also provide a firm footing overland and on rivers too. Thus in January 1853 the Philadelphia, Wilmington & Baltimore Railroad, with its ferry service over the Susquehanna River blocked, laid track almost a mile in length over the ice that had become several feet thick. The company transported all passengers in sleighs, but during its five weeks' use of the ice bridge, employed teams of horses to haul 1378 loaded freight cars across the river to maintain regular service.[16]

---

[15] Ludlum *EAW*, vol. 2, pp. 156–58, quoting the *Chicago Daily Press*, Jan. 23, 1855, and Earl S. Miers, ed., *Lincoln day by day. A chronology, 1809–1865* (New Brunswick: Rutgers University Press, 1961), pp. 136–37; *Milwaukee Sentinel*, Jan. 23, 1855.

[16] Ludlum, *EAW*, vol. 2, pp. 46–47, 224–25; *Rochester Daily Democrat*, Dec. 29, 1852; *Weatherwise* (Dec. 1954): pp. 156–58.

Snow in the Cities

1856 would again spread deep snows and thick ice over a vast area. At Detroit and Milwaukee, where the temperatures dropped to a record −23, and at Chicago to −24, sufficient snows fell to provide good sleighing for two months and more. At both Cincinnati and St. Louis, the ice-clogged rivers that had halted all traffic for several weeks suddenly broke loose on February 24 and 25, sweeping their stranded fleets of steamers and other craft downstream, many to utter ruin. Great crowds, summoned to the scene by tolling fire bells, watched the dramatic spectacles from snow-covered banks. Early in December 1856 Milwaukee received another heavy snow, which gale winds piled into deep drifts that blocked city traffic; the blinding storm obscured the fate of several ships riding at anchor off the docks and of others seeking its port; only one of several trains arrived during the storm.[17]

But it was at New York that the storm inspired the most revealing account of wintertime conditions. Fortunately that city had an observant diarist in George Templeton Strong who was attentive to the experiences and responses of the average citizen. His entry for Tuesday, January 3, 1856, merits reproduction in full:

> This is a stern winter. Saturday's snowstorm was the severest for many years past. The streets are like Jordan, "hard roads to travel." One has to walk warily over the slippery sidewalks and to plunge madly over crossings ankle-deep in snow, in order to get uptown and down, for the city railroads are still impracticable and walking (with all its discomforts) is not so bad as the great crowded sleigh-caravans that have taken the place of the omnibi. These insane vehicles carry each its hundred sufferers, of whom about half have to stand in the wet straw with their feet

---

[17] Ludlum, *EAW*, vol. 2, pp. 156–60, 164, 227–30.

freezing and noses tingling in the bitter wind, their hats always on the point of being blown off. When the chariot stops, they tumble forward, and when it starts again, they tumble backward, and when they arrive at the end of their ride, they commonly land up to their knees in a snowdrift, through which they flounder as best they may, to escape the little fast-trotting vehicles that are coming straight at them. Many of the cross streets are still untraveled by anything on wheels or runners, but in Broadway, the Bowery, and other great thoroughfares, there is an orgasm of locomotion. It's more than a carnival; it's a wintry dionysiaca.[18]

The Introduction of Snow-plows

While the residents of New York and other cities "were busy digging themselves out," as Strong put it in another diary entry, chugging locomotives were valiantly bucking the drifts that periodically blocked the rail lines fanning out from each metropolis and connecting many cities. Most railroads prided themselves on maintaining or quickly restoring service after a storm. Many must have followed the example of the Baltimore & Ohio which, as we have seen, had employed "a contrivance" to remove snow in its first winter. Although specific mentions of the early use of snowplows are missing, two widely-separated inventors secured patents for such devices in 1846, and after the big storms of 1856 a dozen more submitted designs.[19]

---

[18] Allan Nevins and M.H. Thomas, eds. *The Diary of George Templeton Strong, The Turbulent Years, 1850–59*, (New York: Macmillan Company, 1952), pp. 150–51.
[19] Ludlum, *EAW*, vol. 2, p. 100.

The number actually produced and put to use is not known, but scattered evidence supports their wide introduction. Gerald Best, in his fascinating account of snowplows on early western railroads, describes a wedge shaped plow made of sheet iron and attached to the cowcatcher, which he says was the "standard practice on eastern railroads from the early days." Several of the railroads that tried to buck drifts with two or three engines in a line in 1856 may have equipped the lead engine with such wedges, which Best called a "pilot plow". The New York Central, however, brought an "improved snowplow" to Rochester in January 1857 for use on its upstate lines. The new plow, invented in Philadelphia, was mounted on wheels and had a broad shovel scoop to lift the snow from the tracks, and a wedge near the back or top of the tilted scoop to shove the snow off to the side, or onto a movable chain designed to carry it back to a dump car behind. Again the actual use of this rather elaborate plow is undocumented, but the Central used a railroad plow of some sort to clear its tracks near Rochester after a heavy snow in January 1864. It was some two years later that the much sturdier Bucker plow, built in Sacramento on somewhat the same design (but minus the movable chain) made its appearance on the Central Pacific and played a major role, as described by Best, in the building and early operation of the transcontinental lines.[20]

The first use of plows by horse cars in city streets is also uncertain but an early mention comes from Milwaukee in January 1862. It appeared incidentally in a brief item on "The Snow" that captured the light-hearted spirit with which editors of aspiring journals treated local occurrences.[21]

[20] Gerald M. Best, *Snowplow: Clearing Mountain Rails* (Berkeley: 1966), pp. 39–49; *Rochester Daily Democrat*, Feb. 7, 9, 1856; Jan. 10, 1857; *Rochester Union & Advertiser*, Jan. 20, 1864; U.S. Patent Office, *Reports*, (1856) III:275.
[21] *Milwaukee Sentinel*, Jan. 7, 1862.

Sleighbells and Steam Whistles in the Snow

THE SNOW—The snow was used to advantage yesterday. Every description of cutter, sledge, and sleigh was out, and the merry jingle of the bells lent a wintry charm to the day, which properly belonged to Christmas. The officers of the horse railroad worked indefatigably to keep the track clear, and with the assistance of a serviceable snowplow were entirely successful, and the regular trips of the cars have not been interrupted.

Another early mention of the use of snow-plows on horse car lines appeared in a Rochester paper two years later. A stipulation requiring the removal of ice and snow from its tracks had been included in the horse-car company's charter, and its managers hurried to comply on the onset of the first snowstorm after its opening late in 1863. As the snow increased in February the horse-plows became stuck, and the company secured permission from the council to shift to sleighs, as horse-car companies in other cities had frequently done. Three years later, however, when a *New York Times* reporter noted the use of plows on that city's car lines, he characterized it as "prudent action," but did not greet it as a novel procedure. The car companies, by driving snow-plows drawn by four horses over their tracks during the night, had all routes in operation the next day.[22]

As the snow-plows became more prevalent, a reporter for the *Milwaukee Sentinel,* who had read articles describing rides on speeding engines, published an account of his own ride on a railroad snow-plow. Watching calmly through the windows of the heavy car pushing the snow, he had the sensation of standing at the prow of a ship steaming through a white sea casting the snow aside "like foam from

---

[22] *Rochester Union & Advertiser,* Feb. 17, 1864; *New York Times,* Jan. 18, 1867.

its bows." The experience, he declared, "transcended in excitement riding on an engine, as going skyward in a balloon does the quickest ground travel." With a powerful engine behind, it attained a speed of 30 miles an hour, he reported, clearing the track of a ten-inch snow that had drifted in places to a depth of several feet.[23] In Boston, the *Evening Traveller*, after describing the severe character of a storm in January 18, 1867, commended the Boston & Maine, which had despatched its Portland train at the usual hour with four engines and a snowplow to clear the track. Although the trip was slow and tedious, the effort was to be applauded. The *Milwaukee Sentinel* took a similar view a week later in its report of a battle by the Western Union Railroad in Wisconsin to re-open its route with the aid of "superior snowplows." The snow, which reached a depth of two feet that month, halted all traffic in the city and "the horsecars did not attempt to run," but confidence was high that travel would be restored as soon as the storm ceased.[24]

Other indications of a more aggressive response to the urban snow problem came from widely scattered directions. In New York City, a *Times* subscriber wrote on January 5, 1859, to protest the action of the stage coach companies who had scattered salt along their route on Broadway to keep a lane open; the practice, he declared, spoiled sleighing for the public. The editor, in reply, defended the innovation as an economical and practical way of maintaining wheeled traffic on that busy artery. Good sleighing was possible on Fifth Avenue and many other streets, he noted, adding a word of commendation for the car companies, which had battled all night [no mention of plows!] to keep most of the routes open.[25]

---

[23] *Milwaukee Sentinel*, Feb. 27, 1866.
[24] Ludlum, *EAW*, vol. 2, pp. 75, 174–75, quoting the *Evening Traveller*, Jan. 19, 1867, and the *Milwaukee Sentinel*, Jan. 26, 1867.
[25] *New York Times*, Jan. 5, 1859.

Several cities, plagued by increasing fire hazards particularly in the winter season, were reorganizing their fire fighting departments in the late fifties. The volunteer companies, racing one another through the snow to be first at a fire, had frequently resorted to snowballs to impede competition. Their zeal for precedence had often reduced the effectiveness of their inefficient hand pumps. When a manufacturer in Seneca Falls produced a practical steamer, priced at $5,000 in 1858, several cities (including Buffalo, Rochester, and Syracuse) acquired one or more engines and recruited paid companies of mature men to operate them. Soon teams of horses drawing blazing steamers through the streets, with volunteers hauling hose and ladder carts in their wake, added new excitement particularly in snowy weather when the hazards both in the streets and at the fire were magnified.[26]

Concern over snow and ice on the sidewalks was also mounting. In New York, as *The Times* observed, a snowstorm presented opportunities to "thousands of poor people who seek employment and are glad to find it and to earn their half dollar clearing away snow from sidewalks. ...It is not an exaggeration," the editor added, "to say that yesterday, in New York and Brooklyn, some $30,000 were expended in this charity; and" [he concluded somewhat too blithely] "even a snowstorm, pitiless and merciless though it be, still brings a blessing to many a poverty stricken outcast."[27] Together with several other northern cities, Rochester had an early ordinance requiring householders to keep their sidewalks clear of ice and snow. Noncompliance prompted the council to adopt an amendment in 1868 directing the Superintendent of Streets to clear neglected sidewalks after twenty-four hours and to add the cost to the assessments. In New York City where the

---

[26] *Rochester Democrat & American*, June 7 and 18, Aug. 19, 25, 27 and 29, 1858.
[27] *New York Times*, Jan. 5, 1859.

enforcement of such a law was suddenly stepped up during a storm that February, "An Old Citizen" wrote to protest when a police officer's testimony overweighed his own denial of neglect; again the editor supported the official action as necessary to keep the walks open to the public.[28]

A succession of mild winters frequently followed exceptionally severe ones, spacing the blows far enough apart to relax the guard in many cities. A frigid but dry winter could supply a respite, too, welcomed by residents who were eager to get out their skates. That sport had become increasingly popular during the fifties, and a cold season in 1859 and again in 1866 drew thousands of skaters to the lake in Central Park and to ponds and rivers in other northeastern cities. Skaters in New York and Boston enjoyed these interludes to the full, but some inhabitants of these cities, unable to forget that a snow or ice blockade could impose great commercial losses, were seeking new precautions.[29]

A few other cities, such as Buffalo and Rochester, which received heavy snow deposits almost annually, were also becoming alert to the snow hazard and aggressive in their response. A few enacted new building codes—that at Buffalo to safeguard structures against violent winds and waves off the lake as well as against heavy snows. Rochester received warning of the latter danger when a heavy wet snow in April 1857 collapsed its recently constructed suspension bridge over the Genesee Gorge, but its council dallied about reform of the building code until several roofs caved in under a heavy snowfall a decade later. The Western Union Telegraph Company headquartered there was, however, actively pressing the re-

---

[28] *Rochester Revised Charter* (1892), pp. 93, 97; McKelvey, "Snowstorms and Snow Fighting," pp. 3–4; *New York Times*, Feb. 29, 1868.

[29] Ludlum, *EAW*, vol. 2, pp. 63, 69; F.R. Dulles, *America Learns to Play* (New York: D. Appleton-Century Company, 1940), pp. 96–97.

construction of its lines in these years as a safeguard against winter storms.[30]

The Civil War, fought largely in the border states and the Washington area, had less impact than the Revolution on urban responses to snowstorms. Cold waves and harsh weather with some snow did dip down into Maryland and Virginia and even into the lower Mississippi Valley, harassing both armies and obstructing campaigns. But the same operations interrupted the flow of meteorological reports from key points to the Smithsonian Institution and hindered scholarly attempts there to develop a weather warning service. The military campaigns and the continuing efforts to provision the armies also disrupted some commercial supply routes in the North and contributed to the development of a coal famine in several cities during the cold winters of 1863–64 and 1864–65. New York City again experienced a fuel shortage similar to that during the Revolution; this time it was coal, not firewood, that was in short supply, forcing the prices up sharply. Rochester, lacking direct railroad connections with the coal fields, suffered a coal famine too and blamed it in part on neighboring Syracuse and Buffalo where coal trains bound for Rochester were unloaded. Concerted citizen efforts failed to raise sufficient funds to buy a coal mine in Pennsylvania but did produce a Consumer's Benefit Coal Company to purchase a quantity of coal in the summer months to supply business and household needs in the second hard winter.[31]

The *New York Times*, with telegraphic connections throughout the country, now published extended lists of dispatches noting the incidence of snowstorms each winter. One column on January 21, 1867, included reports

---

[30] *Rochester Daily Democrat*, Apr. 22, 1857; *Rochester Union and Advertiser*, Feb. 10, 1865 and Dec. 8, 1868; Ludlum, *EAW*, vol. 2, pp. 61, 172.

[31] Ludlum *EAW*, vol. 2, pp. 121–37; Blake McKelvey, *Rochester The Flower City: 1855–1890* (Cambridge: Harvard University Press, 1949), pp. 73–74.

from twenty-two cities as far apart as Portland, Maine, St. Louis, and Washington, D.C. The *Times* was also ready to supply fuller accounts of severe local storms, such as the one which had hit the city a few days earlier in January. In a chatty two-column article, an observant reporter described the snow-blanketed city. Early risers that morning had found the streets covered to a depth of barely six inches in the center but with high drifts piled against the buildings on both sides. "Milkmen and bakers' boys wandered knee deep in search of their customers.... The car tracks had been wisely cared for by the snowplows during the night.... Luckily for the poor horses, four of them to a car, the number of riders was greatly reduced." Describing Broadway, which "is to tell the story of the City," the writer reported, "As early as noon, the sleighs began to show themselves, of every shape and size. The expressmen seemed to be the first in the field on runners. These were soon followed by the less shapely... sleds of the cartmen, and these again by the gaily-painted equipages of the pleasure-lovers, all gliding into the great avenue from every street." A renewal of the storm, with winds reaching gale proportions, drove many indoors that afternoon, but the storm subsided around nightfall and "by 10 o'clock every street was alive again with the glittering throng and the very air made jocund with their merriment," which elicited another full column from the reporter.[32]

Three years later George Templeton Strong recorded the onset of a new storm on December 28, 1879, with a wry comment: "The evil snow is upon us" again. He, however, indulged in an optimistic speculation, "A century hence cities will be put under glass and New York will be enclosed in a huge crystal palace."[33]

But if some eastern and Great Lakes cities were developing new responses to winter storms, new towns beyond the

[32] *New York Times*, Jan. 18, 1867.
[33] Nevins and Thomas, *The Diary of George T. Strong*, vol. 4, p. 336.

Sleighbells and Steam Whistles in the Snow

Mississippi were reeling in the mid- and late-sixties under the harsh blows of a succession of arctic waves, heavy snows, and violent winds. With temperatures dropping to −24 at Iowa Falls and to −35 at St. Paul on New Years Day in 1864, and with snows drifting to depths of eight and ten feet at various observations points in 1867 and again in 1870, local observers began to take a more dramatic view of the unexpected rigors. Thus a controversy between the editors of two Iowa weeklies, the *Upper Des Moines* and the *Estherville Vindicator*, over the violence of a storm there in April 1870 first brought the word "blizzard" into use. The word's descriptive vigor quickly won its application in retrospect to many earlier storms and gave a fresh American vitality to winter weather reports.[34]

---

[34] Ludlum *EAW*, vol. 2, pp. 171–72, 175–77, 206–41. The word *blizzard* had an earlier origin, according to *A Dictionary of Americanisms*, but its popular use apparently dates from the Iowa blizzard.

ALBANY'S SNOW FORT UNDER ATTACK
(Courtesy of The New York State Historical Association, Cooperstown)

The snowball attack on "Fort Orange" in January 1888 was one of the most spectacular of several winter pageants in an era when a heavy snow was a challenge, not a threat to community spirit.

CHAPTER THREE

STORM WARNINGS, SNOW PLOWS AND BLIZZARDS

1870–1890

The supine acceptance of blustery snowstorms as fateful though sometimes benign wintertime interludes was beginning to wane in a few cities by the late sixties. Impatience for the prompt delivery of the mails, essential produce, and other urgent supplies, joined with the old concern for the safety of travelers (now including mariners on the Great Lakes) to spur efforts to reduce or at least mitigate these hazards. Fortunately the widespread network of weather watchers, reporting by telegraph to the Smithsonian Institution, was enabling analysts there to chart the course of successive storms and suggested the feasibility of issuing warnings of threatening conditions. Railroads, struggling to maintain service in the increasingly competitive inter-city trade, acquired new and larger plows to buck wintertime drifts; horse car companies, eager to collect fares from the many pedestrians who sought their services in stormy weather, resorted to plows and other devices to clear their tracks; retail merchants, impatient, indeed frantic, when snow mounds blocked access to their wares, demanded that the commercial roads be kept open. But these and similar efforts to combat snowstorms created new hazards for the many urban residents who looked forward to wintertime sleighing and stirred new controversies over the use of salt and the abuse of horses. The mounting conflicts over policy, raging in New York and a few other, cities would disappear however with the onset of the great blizzard of 1888, which reasserted the awesome power of nature and revealed the need for new technological advances.

Weather Forecasting Commences

Among the 150 or so weather-watchers who wired reports to the Smithsonian Institution in the late 1860s were several who wished to make more positive use of their climatological observations. Frank Armstrong, in charge of the weather dispatches for the Western Union at New York, made a copper plate for a standard weather map on which isobar lines could be traced showing changing barometric conditions each day, but he failed to persuade the company to take it up. Professor Cleveland Abbe, director of the Cincinnati Observatory, secured the backing of that city's Chamber of Commerce for a plan to issue daily weather reports in the fall of 1869 and produced his first weather bulletin on September 22. At Milwaukee, battered by severe storms in January 1867 and by heavy snows again two years later, Professor Increase A. Lapham persuaded Congressman H. E. Paine of Wisconsin to promote the establishment of a national weather service to supply warnings of impending lake storms. The U.S. Signal Service, reorganized in the mid-sixties under the direction of Colonel A. J. Myer, welcomed the new assignment as an expansion of its functions and placed Lapham in charge of the signal service for the Great Lakes district with his headquarters at Chicago.[1]

---

[1] *New York Times*, Mar. 21, 1885; *Monthly Weather Review*, Jan. 1931, pp. 69–70; Patrick Hughes, *A Century of Weather Service: 1870–1970* (New York: Gordon & Breach, 1970), pp. 5–6, 21–24; Ludlum, *EAW*, vol. 2, pp. 174, 210. These developments are discussed in fullest detail by Donald R. Whitnah in his *History of the United States Weather Bureau* (Urbana: University of Illinois Press, 1961), pp. 1–23.

Lapham issued his first storm warning on November 9, 1870, basing it on reports from twenty-four stations. His first bulletin told of high winds at Omaha extending eastward to Chicago and Milwaukee, with falling barometer readings in Detroit and as far east as Rochester. In each of these and other Great Lakes cities, local observers attached arrows and isobar lines to a weather map displayed at the local board of trade or in some other prominent location. In Rochester the map was hung in the Reynolds Arcade where citizens congregated to pick up their mail, send and receive telegrams, consult the reference library, and transact other affairs. Curiosity concerning the newly instituted weather maps, on which the arrow and pressure lines were attached daily in conformity with the latest bulletins, prompted some citizens to push the symbols to see if they were automatic or artificial. A few visiting husbandmen as well as merchants climbed the stairs to the weather observer's office in quest of more explicit forecasts.[2]

Col. Myer had meanwhile engaged Professor Abbe as chief assistant at a newly opened Washington office, which became the national headquarters of the Signal Service in 1871. By vigorous action they increased the number of reporting stations within a year to fifty-five, scattered from Portland, Maine, to San Francisco in the West, and to several Gulf ports in the South. When Abbe issued his first "Weather Synopsis and Probabilities" on February 19, its more cautious character set the pattern for later bulletins, which superseded Lapham's warnings after the great Chicago fire disrupted the service there and speeded Lapham's retirement. Although devoid of the eye-catching arrows and isobars, Abbe's synopses won frequent publication in the *New York Times* and in leading dailies in other cities. In June 1872, the U.S. Congress extended the Signal

---

[2] Hughes, *A Century of Weather Service*, pp. 20–21; *Journal of the Rochester Weather Station*, Nov. 16–Dec. 12, 1870.

Service to cover all the states, not simply those with port cities. Agricultural and commercial concerns inland received increased attention; and to enhance the coverage, meteorological information was exchanged with a service in Canada.[3]

Heightened interest in the weather prompted commercial groups in several cities to contribute funds for the erection of a flagstaff for the display of daily weather signals. The Reynolds Arcade in Rochester donated use of the flagpole on top of its cupola to the local weather observer. Observant citizens learned to recognize the display of a white flag above a yellow flag as the promise of a fair day, a yellow flag above a blue flag as the sign of a rainy day, and a white flag with a large blue sunspot on it as warning of snow.[4] In New York City, where in February 1870 the Signal Service had established one of its first local stations in the Board of Underwriters Building at 49 Wall Street, the office had moved a few months later to 120 Broadway where the Equitable Life Insurance Building afforded a more lofty base for its observations. Its location assured business and civic leaders easy access to its daily bulletins based on periodic meteorological observations made in the heart of the commercial district, and those recorded at a weather station maintained by the city in Central Park, as well as those received by telegraph from Washington. Warnings of an approaching storm on January 26, 1871, prompted New York's horse car companies to engage extra teams to operate their snow-plows throughout the night and enabled them to maintain regular service the next day despite a fall of nine inches.

---

[3] Whitnah, *History of U.S. Weather Bureau*, pp. 22–28.

[4] A few years later a somewhat more complicated series of flag symbols was adopted a the Weather Bureau standard. Many city Directories published a chart of flag signals and explanation. See: *Rochester Directories* (1896–1925), p. 4 or 6 in most issues.

Sleighing parties found ideal conditions for jolly drives and spirited racing in Central Park.[5]

Early Snow-Fighting Dilemmas

In Rochester, however, the vigorous efforts of its horse car company in clearing its tracks during a twelve-inch snow that February plowed uneven mounds into the path of frustrated sleigh drivers. Responding to the protests, the street superintendent hired forty men to shovel it back onto the tracks. The workmen, helped by citizen volunteers, could not keep up with the plows, but as neither party would yield, the battle effectively obstructed all travel in the central district and prompted the council to order the company to substitute sleighs for cars during heavy storms. Four years later, when a similar battle erupted in Buffalo between horse car plows and inconvenienced citizens, a Rochester editor congratulated his readers on their happier solution of the problem.[6]

A small city could resort to sleighs or wait out a severe storm. Many workers (as well as school children) welcomed an extra holiday or two between Thanksgiving, Christmas, and the Fourth of July. Even snow shoveling could break the monotony of the six-day work schedule.

But New York City had a flow of pressing business that could not easily be halted or diverted. Even a mere three-and-one-half-inch fall in December 1871 revealed the extent and gravity of the problem. When the horse car companies tried to maintain their full schedules in order to take advantage of the swollen number of riders, Henry

---

[5] *New York Times*, Jan. 27, 1871; *Rochester Union and Advertiser*, Dec. 29, 1870; Whitnah, *History of U.S. Weather Bureau*, pp. 28–30, James K. McQuire, "History of the Weather Bureau Office in New York City," *Weatherwise*, Apr. 1961, pp. 50–52, 71.

[6] *Rochester Union & Advertiser*, Feb. 13, 1871; Mar. 6, 1875; *Buffalo Morning Express*, Mar. 4 and 5, 1875.

Snow in the Cities

Bergh, founder of the Society for the Prevention of Cruelty to Animals, halted several overloaded cars and forced the companies to supply double teams to pull them.[7] A year later a heavier storm, accompanied by the worst gales in two decades, clogged the streets with two feet of snow and "prodigious" drifts, as George Templeton Strong described the situation. The companies hitched double teams to their plows as well as their cars, and the partially-elevated Greenwich Railroad used one of its steam engines to plow its surface route, but the service was poor. The only solution, declared the editor of the *Times*, was to speed the construction and extension of elevated steam roads to surmount the drifts, or to build an underground line from the City Hall north to Central Park.[8]

But the "long remembered" winter of 1872–73 would not wait such a solution. A partial thaw early in January failed to remove the Christmas Day snow, and renewed freezing temperatures coated the streets with ice and brought many telegraph wires and poles down. Lists of accidents, such as pedestrians and horses slipping on the ice and suffering broken legs (and in the latter case having to be shot and carted away), became regular features of successive storm reports. When word reached the East Coast on January 24 of a ten-inch fall of snow in Chicago (where traffic was kept moving by hitching double teams to horse plows and cars by running extra trains on suburban roads kept open by engine plows), New Yorkers braced for a renewed blast. The new storm hit as predicted three days later, but the car companies were ready and determined and managed to

---

[7] *New York Times*, Dec. 20, 1871; Dec. 27, 1872: Nevins & Thomas, *Strong Diary, Post War Yeats*, p. 165; Zulma Steele, *Angel in Top Hat* (New York: 1942) pp. 64–65.

[8] *New York Times*, Jan. 6, 24 and 28, 1873; Feb. 3, 7, 8 and 26, 1874. Three New Yorkers had secured patents for horse-drawn rotary sweepers in the early 1860s, and street cleaning departments in that city and Philadelphia had each acquired some, but this, to the author's knowledge, is the first instance of their use in snow-fighting.

keep their tracks open (using plows drawn in some cases by eight horses) and they successfully maintained service with double teams on most lines. Warmer and happier days arrived, but they in turn passed, and in February 1874, New York and other northeastern cities faced a new snow emergency. The snowfall this time was not so intense and blustery, but while its protracted character enabled the car companies to maintain service, the accumulation was appalling. By a frequent use of horse-plows and newly introduced rotary sweepers, the companies piled up huge mounds or ridges bordering their tracks, creating great hazards for all who tried to cross the streets or to board the cars. Squads of shovelers opened passageways through these ridges, thus further obstructing the adjoining roadbed for horse sleds and sleighs. The latter, however, again enjoyed ideal conditions in Central Park and on streets free of public transit. Reports from Baltimore and Boston, where the snow scarcely reached six inches, told of the eagerness with which hundreds of unemployed men accepted modest sums for shoveling sidewalks and crossways in that depression year. Thousands of poor men found work shoveling sidewalks in New York, too, but a *Times* reporter remarked sardonically, "It is not of course presumed that the street cleaning bureau will show any signs of life" during the emergency. That prediction held fast, and so did the cold spell; two months later another fall of snow stalled even cars with double teams. Only the Greenwich line on its elevated tracks supplied uninterrupted service.

The losses, however, were great, and something had to be done. A new ordinance empowered the street-cleaning bureau of the police department to open the streets during a snow emergency yet provided no additional funds. When a heavy snow arrived in December 1874, the car companies supplied double teams under Henry Bergh's watchful supervision, and sleighs appeared in abundance in Central Park and on many avenues, but the Board of Police Commissioners failed to agree on a proper course of action. The regular street-cleaning force helped to clear a lane for the omnibuses on Broadway and shoveled a number of crosswalks, but the lack of funds inhibited further efforts; only a shift in the weather cleared the streets. A second storm in January brought good sleighing and again slowed wheeled traffic for a few days; fortunately an early return of mild weather enabled New Yorkers to forget their concerns as they read of harsher storms in Kansas and Minnesota where newly-improved railroad plows pushed by two or three engines battled to maintain train service between Kansas City, St. Paul, and other snow-bound cities. St. Louis, Buffalo, and Boston also received paralyzing blows in March 1875, yet none reported new measures to combat the drifts.[9]

The hard winter of 1874–75 would not be forgotten, however. A group of New York merchants sued the 23rd Street Car Company for damages. The company's plows had pushed a ridge of snow and ice in front of their properties, seriously curtailing business, the merchants claimed. When the case finally reached the State Supreme Court, the judge held for the plaintiffs. Faced with a snow emergency, the company had a right to plow snow off its tracks to maintain services, but it was obliged, the judge declared, to remove it from the streets "as quickly as reasonably possible."[10] Fortunately the mild winter of 1875–76 afforded time for officials and transit companies in New York and elsewhere to mull over their new responsibilities.

The respite came to a sudden end at Rochester on December 29, 1876, and at New York four days later. A fall of three feet of snow at Rochester blockaded all streets, halted trains and other traffic, and absolved officials from immediate action. In New York, which suffered a lesser

---

[9] *New York Times*, Dec. 24 and 28, 1874; Jan. 19 and 24, Feb. 5 and 25, Mar. 4 and 5, 1875; *Buffalo Morning Express*, Mar. 4, 1875.

[10] *New York Times*, Jan. 16, 1876.

Storm Warnings, Snow Plows and Blizzards

blow, the horse car companies, with ample warning, got their plows out early and by employing double teams maintained fair service. And as the snow continued, reaching a level fifteen inches, with huge drifts in many side streets, the police commissioners met to consider appropriate action. As the day progressed, small gangs of city workmen began to clear the crosswalks along Broadway and by nightfall the city's first contingent of horseplows, followed by carts and shovelers, began the task of re-opening the omnibus route on that principal artery. As the night wore on, the Belt railroad, the Third Avenue, and other car lines mustered carts and shovelers to follow their plows as they for the first time accepted responsibility for removing the snow in order to maintain their lucrative services. The cartmen dumped thousands of loads into the East River during the night. Sleigh riders and pedestrians as well as patrons of the lines rejoiced the next morning over the more responsible procedures reflected in the improved condition of the streets.[11]

Protests appeared, however, against a renewed use of salt. An earlier ban against its distribution on streets had allowed a limited application on switches and at sharp curves to keep the track operative. But several companies were now scattering it more plentifully along their routes in order to speed the progress of plows and shovelers. Again Henry Bergh protested the cruelty to horses forced to plod through the briney mush in the car lanes. Pedestrians and passengers complained of the damage to their shoes and clothing from slush covering the crosswalks. The New York police sometimes assisted Bergh's efforts to enforce the ban on salt, but the supervision there was not as effective as at Boston where sleigh owners and pedestrians combined to discourage its use.[12]

---

[11] *Rochester Union & Advertiser*, December 30, 1876; *New York Times*, Jan. 3 and 4, 1877.
[12] *New York Times*, Jan. 2 and 3, 1877; Steele, *Angel in Top Hat*, pp. 69–70.

An urgent press of business denied most New Yorkers the pleasure enjoyed at Washington when an unusual fall of ten inches in January 1877 brought out hundreds of sleighs and prompted many owners of carriages to mount them on improvised runners for drives along the Capital's broad avenues. Thousands of men and boys found jobs shoveling sidewalks, and one horse car line maintained service by a vigorous use of its newly acquired plows. In Boston, where the snow that week was not quite so deep, several car companies re-opened their routes by the same means, but in Hartford, with a fall of two feet, and Rochester, where renewed deposits kept the depth at three feet for several weeks, efforts to clear the car tracks had to be abandoned. The city of Rochester employed extra men to help shovel snow back into the trenches of a few lines that had been partly plowed at its order; sleighs and horse sleds displaced the cars and all wheeled traffic, and many workers as well as school children welcomed the extra holiday or two.[13]

But New Yorkers could not stop the working-day calendar. When a twelve-inch fall blanketed the city in January 1879, the car companies brought out plows drawn by six and even eight horses and attached double teams to all cars in a determined effort to maintain service. After the plows had opened the tracks, many horses were hitched to carts and, accompanied by squads of shovelers, cleared the piles of snow from some of the more congested sections. City workmen again helped to keep the omnibuses running on lower Broadway by clearing the crosswalks and loading strips. That widespread storm dropped a foot of snow on Syracuse, where residents switched to sleighs, but in Buffalo and Rochester the horse car companies tackled six- and eight-inch deposits with plows and kept their tracks open despite renewed protests from sleigh drivers.

[13] *New York Times*, Jan. 3 and 4, 1877; *Rochester Democrat & Chronicle*, Jan. 3, 9 and 12, 1877.

Even a six-inch snow commanded forthright action in New York where the *New York Graphic* began publication of a daily weather map that winter. An "old-fashioned snow storm" on December 22 assured a "White Christmas." It also prompted Captain Williams of the police department to marshall a force of shovelers and carts to clear lower Broadway for the throngs of last-minute shoppers. The horse car companies also brought out their plows, rotary sweepers, and carts, and with the city forces increased the total number of loads dumped into the East River to over 4,000.[14]

Municipal Responsibilities Discovered

Despite these accomplishments, New Yorkers felt a need for more responsible action. A reorganization of city government in 1881 established a separate Department of Street Cleaning under the direction of James S. Coleman with a year-round force at his command. Snow removal was only one of his many duties, but his men carted a total of 23,174 loads of it in the winter of 1881–82.[15] The heaviest snows that winter fell in the Midwest, blockading Milwaukee and St. Paul among other places, and downing many telegraph lines. Both of these cities relied on their horse car companies to clear their tracks in order to resume service, but Milwaukee kept its street-cleaning force busy opening the sewer drains and removing fallen wires and limbs. Despite the widespread disruption of telegraph lines, the Signal Service managed to continue its daily dispatches of "Synopses and Probabilities." The early bulletins

---

[14] *New York Times*, Jan. 17, Dec. 23, 1879; *Rochester Union & Advertiser*, Dec. 9, p. 187; *Rochester Democrat & Chronicle*, Jan. 6, 1879; Hughes, *A Century of Weather Service*, p. 192.

[15] George E. Waring, *Street Cleaning*, pp. 1–10, 94–96.

in February 1882 enabled the New York car companies and Commissioner Coleman to get their respective plows, sweepers, shovelers, and carts out for an overnight battle that presented residents with an open Broadway and full service on all car lines by late morning. The "good skating" that had previously attracted hundreds to the lake in Central Park had ended, but sleighs made their appearance in abundance there and on many streets without tracks.[16]

Sleighs replaced the horse cars in Rochester again in December 1882 and also the next February on the suburban horse car line of the Shore Railroad on Long Island. New York however, was developing a more effective alternative; elevated steam railroads that overrode the highest drifts. Its first, on Greenwich Street, which had been extended out along Ninth Avenue, had provided the only continuous service during severe storms for a dozen years. Three new north-south Elevated lines were brought into operation by the early eighties when the long projected Brooklyn Bridge, spanning the often ice-choked East River, was finally opened. But the city's population and that of Brooklyn were growing so rapidly that the new facilities scarcely diminished the number of horse car riders. The old battle to maintain service started again with the arrival of each storm, as on January 11, 1883, and again on December 20. Two new features appeared in the reports on the latter occasion—the despatch of sand carts to coat the approaches to the recently opened bridge and other slippery places, and the employ of squads of newly arrived Italians as shovelers. Coleman despatched 200 of these workmen, with cartmen who could speak English as well as Italian, to open a lane for the omnibus on Broadway and doubled the quantity of snow removed. Many more of these poor immigrants found welcome jobs shoveling side-

---

[16] *New York Times*, Mar. 12 and 13, 1881; Jan. 6 and Feb. 6, 1882; *Rochester Union & Advertiser*, January 22, 1881; *Milwaukee Sentinel*, Mar. 13, 1881.

walks to clear the fashionable avenues for another White Christmas that year.[17]

A pair of two very cold but relatively dry winters in the Northeast gave New York a respite from snow-fighting and encouraged a revival of wintertime sports. An arctic wave, which started by dumping the "heaviest snow in memory" on Portland and other parts of Oregon in the middle of December 1884, halted all traffic in that city and on railroads heading east as far as Kansas City and Minneapolis. With huge plows the railroads broke the blockade there, but as the cold wave moved east, dropping temperatures to record lows (-40 at Utica), it coated rivers and ponds with a thick crust of ice that brought out skaters in abundance on many urban widewaters. An ice regatta, first staged by three iceboat clubs on the Hudson during the frigid February of 1875, was repeated there and at other ice-covered expanses in succeeding cold winters. In January 1885 two clubs of ice yachtsmen vied for preference on the Hudson, and already the sport had spread upstate and into New England. In Rochester a yachtsman constructed an ice yacht, named the *Zephyr*, which shot ahead at such speed on its trial run over the frozen surface of Irondequoit Bay that it scattered its inexperienced crew along its course and crashed at Stony Point. Thousands of skaters converged on the lake in New York's Central Park after a sudden freeze on December 21 for a day of exciting sport which, however, ended as rising temperatures transformed the light snow into rain that evening. A second cold wave a few weeks later restored skating in the park and brought a season of good sleighing there and on several track-free avenues. Bobsled clubs had appeared in several upstate cities in the early eighties and, as the fad spread, a number of them congregated in 1886 for races

---

[17] *Rochester Union & Advertiser*, Dec. 4, 1882; *New York Times*, Jan. 11 and 12, Feb. 12, Dec. 20 and 26, 1883; Waring, *Street Cleaning*, p. 95.

on a steeply sloped avenue in Albany. They gathered there again in January 1888 when the city staged a winter carnival in one of its parks, featuring an ice castle named Fort Orange, which was attacked and, assisted by a shift in the weather, demolished by thousands of snowballers in a joyous climax on February 2.[18]

While enjoying their good fortune, New Yorkers read of harsher conditions in the West. At St. Louis, Milwaukee, and Chicago, snowfalls of two and more feet in some places halted street traffic for a day or two, blockaded railroads, and felled telegraph lines as far east at Buffalo. Horse-plows, rotary sweepers, and horse cars with double teams restored service in the western cities, and steam-plows re-opened their connecting railroads, but a second hard winter brought renewed punishment there.[19] The big storm of February 1886, however, hit with greatest severity at Louisville, Washington and Baltimore, clogging their streets with fifteen inches or more and halting all wheeled traffic for two or three days. It was the first major storm since the start of national weather records to scourge the southern fringe of the snow country and thus helped to determine that boundary. Most cities farther north escaped its fury. New York, with a modest five-inch fall, had its street-cleaning crews out—500 men with 400 horses and carts—and kept the streets in the commercial district open throughout the storm. On affluent Fifth Avenue the omnibus company substituted huge sleighs for its deluxe carriages and supplied jolly rides at the regular fares for passengers limited to twenty-five a load. Citizens who wished to reach their destinations more quickly trudged through side streets and climbed the

---

[18] *New York Times*, Dec. 17, 21 and 22, 1884; Jan. 18, 1885; Mar. 7, 1896; 15–1; *Rochester Democrat & Chronicle*, Feb. 19, 1900; "Albany's Winter Carnival," *Heritage* VI, no. 3 (Jan.-Feb. 1990).

[19] *New York Times*, January 17, 18 and 19, 1885.

stairs to the elevated platforms where steam trains speeded north-south traffic.[20]

Mounds of snow were not the only hazards confronting those in the cities during the storms of the early eighties. Broken and fallen telegraph lines had long inflicted heavy financial losses and had disrupted communications, but the tangled mess of telegraph, telephone, and electric wires that littered the streets of Chicago, Milwaukee, and Washington as well as New York during several of these storms, posed serious safety threats. A mounting demand for their removal into underground conduits or tunnels produced legislative action in New York and prompted the drafting of similar measures in other states. The National Electric Light Association debated the question at successive meetings, protesting the huge cost involved, especially in view of the unsatisfactory performance of the few early experiments with underground installations in Boston, Milwaukee, and Philadelphia. But the periodic losses to the companies from downed wires, as well as the occasional fatalities that resulted, spurred the search for a safe and enduring insulation and for an economical system of underground installation to meet the deadlines posed by legislation.[21]

Among the groups pressing for underground conduits were the police and fire departments of snow-battered cities. Not only were their responsibilities increased by the hazard of fallen wires, but the tangled mesh of wires frequently included some from their own communication systems, which were thus broken in a crisis period. Most of the major cities had installed electric fire alarm and police telegraph systems before the 1880s, when Chicago, Milwaukee, and Brooklyn first introduced telephone systems

---

[20] *New York Times*, Feb. 4, 5 and 6, 1886.
[21] *Scientific American*, Sept. 25, 1886, p. 192; May 26, 1888, p. 326; Sept. 8, 1888, p. 144; Feb. 7, 1891, p. 86; *Scientific American, Supplement*, Sept. 26, 1891, pp. 13116–17.

Snow in the Cities

as well; heavy sleet and wind storms not only inflicted losses but also reduced effective cooperation among their scattered stations. The simultaneous blackout of the newly introduced electric street lights in several snow-bound cities in the mid-eighties added to the demand for underground conduits.[22]

A moderately hard winter in 1886–87 dropped a near record total of forty-nine and one-half inches on New York, but the snow fell in well-spaced storms that enabled the plows and the cartmen to perform their tasks effectively. Protests against the city's practice of despatching workmen to open the omnibus lane on Broadway had finally led to an agreement by the newly consolidated Metropolitan Street-Railway Company to contract with the Holland Company to perform that function. Relieved of this burden, Commissioner Coleman's forces, together with the contractors, engaged to open the roads south of Fourteenth Street, carted a total of 64,408 loads to the East River docks without exceeding his $25,000 appropriation. The frequent snowfalls that winter kept sleigh riders in a happy mood. When a blustery storm hit early the next January, clogging traffic while the fair-weather flags were still flying, many observers were content to poke fun at the weather service.[23]

The Signal Service had acquired a degree of bureaucratic self-confidence by the mid-eighties. It had extended its network of observation stations throughout the country, and many continued to display warning signals from flagstaffs mounted on top of lofty towers. The eight-story Powers Block in Rochester, with a tower rising another thirty-six feet, now supplied the appropriate pedestal. A timely warning of a storm there on January 10, 1886, had

---

[22] U.S. Bureau of Census, *Telephones and Telegraphs, 1902* (Washington: 1906), pp. 124–28, 146; ibid, *Statistics of Cities (1905)*, p. 354; Bessie L. Pierce, *A History of Chicago* (New York: Alfred A. Knopf, 1960) pp. 329–30; *New York Times*, Jan. 26 and 27, 1891.

[23] Waring, *Street Cleaning*, pp. 94–95; *New York Times*, Jan. 14 and 27, 1888. Whitnah, *History of U.S. Weather Bureau*, p. 57

enabled merchants to take necessary precautions and brought an expression of thanks from a local editor. The weather analysts of the Signal Service in Washington seldom heard either the favorable or the unfavorable comments. They were busy charting the scattered reports of temperature, pressure, and rain or snow by isobar lines and other symbols on comprehensive maps that disclosed conditions and trends with great objectivity. Their objectivity had, however, acquired a degree of detachment that sometimes insulated them from threatening possibilities, as it did early in March 1888. Elias B. Dunn, chief of the weather station in New York, receiving no cautionary warnings from Washington on the 10th, issued a prediction, based partly on local conditions—"cloudy, followed by light rain and clearing" for Sunday the 11th. His first inkling of more serious developments came when he returned to his office through a fairly heavy rain on Sunday afternoon only to find the expected 5:00 P.M. message from Washington missing and all telegraph connections with the capital severed.[24]

The Blizzard of 1888

A storm center, first noted in the North Pacific on March 6, 1888, had stirred little concern. In charting its course across the continent, the analysts noted that it assumed on Sunday the 11th a somewhat unusual form as a trough of low pressure extending from a northern center swirling counter-clockwise over Lake Superior to a southern center of diminishing strength over Georgia. As the northern center raced eastward, drawing frigid air from Canada into New York and New England, the southern center

[24] Whitnah, *History of U.S. Weather Bureau*, p. 24; *Rochester Union & Advertiser*, Jan. 11, 1886, 2–3; Irving Werstein, *The Blizzard of '88* (Ithaca: Cornell University Press, 1960), pp. 9–17.

rushed north to New Jersey, increasing the wind velocity, pushing moisture into the path of the arctic stream from the north, and transforming the rain clouds over New York on the 11th into a blinding snowstorm on the 12th. When "Farmer" Dunn, as he was familiarly known, reached his office with difficulty on Monday morning and received only fifteen of the accustomed 175 reports, he quickly realized that the advancing storm had already downed telegraph lines over a wide area and had left New York and probably many other northeastern cities isolated. It was another day before a cable from London advised him of the similar plight of Boston whose only outside contact was by underwater cable to England.[25]

If New York's weather eyes were half shut and blinking, those of its officials and most residents were practically blinded by the unprecedented blizzard. Many early risers on the 12th were dazed to find their doors and first-story windows covered by packed snow, only to look out from an upper floor and see the pavement across the street bare. Vigorous men who tunneled or walked out, as their locations warranted, faced blustering winds that piled the falling snow into great mounds and ridges. Those with urgent duties searched in vain for horse-cars or hacks, many of which had already been left stranded and half covered in the drifts. Many milk carriers, coal dealers, and other hucksters who had started out in pre-dawn hours had also become marooned as the snow increased. Determined businessmen, clerks, and other workers plodded and staggered to the stations of the elevated lines to board one of the steam trains that had proved a mainstay in earlier storms. Soon the station platforms were crowded with shivering commuters who quickly jammed all cars as they arrived and so overloaded the trains that the puffing

---

[25] Strong, *The Great Blizzard of 1888* pp. 5–6, 11–12, quoting later statements by "Farmer" Dunn and a report by General A. W. Greely, Chief Signal Officer of the Army.

engines had difficulty pulling them along the snow-packed trestles. Several trains became stalled on one line and had to unload passengers by ladders supplied by fire companies or brought from nearby barns by men, who in some cases exacted silver coins for the service. In one frightening case a loaded train that had passed a crowded station without stopping, in order to avoid additional delays, plowed into the rear of the overloaded train struggling to get up steam ahead of it, killing one of the engineers and injuring a score of passengers in the crush. By nightfall even the elevateds had stopped running, and except for the wind, New York, the second biggest city in the world, was as quiet as a tomb.[26]

New York City, with over two million inhabitants, was paralyzed. Four-fifths of its 10,000 telephones were silent, and practically all of its electric lights and the great majority of its more numerous gas lamps were blacked out that night. Residents who had braved the storm early in the day had either turned back quickly or found themselves in almost deserted offices or shops. By nightfall, an estimated twenty inches of snow had fallen. "The only lights to be seen up and down the dreary waste of snow," an intrepid writer for the *Herald* reported the next day, "were those that shone from the comfortable interiors of several saloons that are scattered along Broadway, and it is needless to say that they were well filled with the blizzard-filled people. ...It is estimated that the cost to New York of this frolicsome little storm will aggregate millions of dollars....The hotels, however, reaped a harvest... as every room, nook and corner...was crowded to the utmost."

Over 400 were later found buried in the drifts or otherwise counted as victims of the storm. Among those who died from natural causes that day was Henry Bergh who

---

[26] Strong, *The Great Blizzard of 1888*, pp. 7–10, 79 quoting among others the *New York World*, Mar. 13, 1888; *New York Times*, Mar. 13, 1888; Werstein, *The Blizzard of '88*, pp. 34–56, and passim.

was thus saved the painful sight of the many horses driven to impossible tasks during the early hours of New York's great blizzard.[27]

The breakdown in communications was appalling, and yet one of the few services maintained throughout the storm was the daily press. "The *World's* big city department was run yesterday on the plan of an exploring expedition," its editor declared. "Rumors of accidents and incidents poured in...and were run down by the speediest methods; ...hundreds of miles were covered on foot...and today you have the news." The *Times* was similarly busy and, with its telegraphic sources cut off, filled four pages with accounts of the unsuccessful efforts of the car lines to open their routes, of the halting of car service and pedestrian passage over Brooklyn Bridge, and of the gradual suspension of ferry service; it also printed descriptions of the tangle of wires brought down by the storm and of the crowds lining the platforms of elevated lines that could not keep their trains moving; its editor called again for the construction of undergrounds, both for safe transit and for uninterrupted communications. Such proposals were of course for the future, and in the meantime, enterprising boys were required and found to haul bundles of papers to distributions points in hotels and saloons throughout Manhattan. Two young "huskies" dragged a sled load of New York papers, including "Blizzard Editions" of the *Sun* and the *World*, across to Passaic the next day.[28]

Prominent among the services that continued to function in the face of great difficulties was the fire department. A member of Engine Company 33 recalled years

---

[27] Strong, *The Great Blizzard of 1888*, pp. 7–8, 19; *New York Times*, Mar. 13, 1888.

[28] Strong, *The Great Blizzard of 1888*, pp. 8, 22, 28, 66; *New York Times*, 13 and 14, 1888; Werstein, *The Blizzard of '88*, pp. 124–26; See also, Eugene Kinkead, "New York City Weather Extremes," *The New Yorker*, Jan. 31, 1977, pp. 59–60; *American Weather Stories*, pp. 58–75.

Storm Warnings, Snow Plows and Blizzards

later how his double company had hitched its two teams to one engine and had responded to two emergency calls on Broadway. On racing to the first fire, with several of the men leading and breaking the way for the horses and the rest following with the hose cart, they had passed another engine company stuck in a drift, but could not pause to free it. At both of their fires they were fortunate in finding convenient hydrants on wind-swept corners and extinguished the blaze without difficulty. "The Fire Department," as he recalled, "was very lucky,...there were only a few fires," and most of them were quickly reached by the widely scattered engine companies.[29]

Some essential services, however, had to be suspended. The schools, of course, remained closed for several days, and most public meetings were deferred. And, as the supreme measure of a great storm, funerals were postponed and the corpses of deceased loved ones had to be kept in snow banks in the city or in frigid barns, as in Hartford.[30]

Despite their huge plows, the major railroads serving the city were effectively blocked. An early commuter train on the New York Central and Hudson line had become stalled when trying to push through a thirty-foot drift in the deep Spuyten Duyvil Cut in the Bronx a few miles north of the city. Other trains loaded with impatient commuters, and trains with sleepers from upstate, had quickly lined up behind it, until a string of eight trains unable to move forward or backward became marooned in snow. Fortunately the telegraph operator at "Spike," as the cut was familiarly known, took the initiative in organizing a food service. Enlisting the aid of a nearby grocer and butcher, he supplied sandwiches and hot coffee to the passengers imprisoned in the mile-long string of cars for

---

[29] Strong, *The Great Blizzard of 1888*, pp. 20, 66.

[30] *Ibid.*, pp. 40, 56; See also Nat Brandt, "The Great Blizzard of '88," *American Heritage* (Feb. 1977): pp. 33–41.

two days, until a huge plow pushed by three engines, and assisted by a contingent of shovelers, opened the cut.[31]

When the gale subsided and the snow slackened on the 13th, shovelers by the thousands tackled the job of digging out. One merchant, who had bought a car load of snow shovels at end-of-season prices a week before, quickly disposed of his stock with a handsome profit. So many shovelers found jobs breaking paths into homes and shops and along the sidewalks that the horse car companies had difficulty recruiting men to clear their tracks. Several companies operated sleighs for a day or two while even the elevated lines remained closed. The street-cleaning forces tackled the job of opening 23rd Street and other streets leading to the East River docks, thus providing access for the hundreds of dump carts that worked through the night clearing Broadway and other north-south avenues. Although a few business men, unable to reach their destinations on the 13th, were heard to grumble about the city's inefficiency, most took their hardships with good humor. The next day, as the cars started moving again, passengers exchanged boisterous accounts of their ordeals, and Coleman received commendation for his efforts, which had boosted the number of cart loads of snow dumped into the river to a new high, 76,100, not counting the still greater quantities removed by the car companies. Best of all, the city was again in business and making up for lost time.[32]

As New York freed itself, news of hardships elsewhere began to filter in. The Great Blizzard had been widespread. Philadelphia had not been as severely battered, but a snow blockade had halted all wheeled traffic until the car company had mustered an army of shovelers to assist

---

[31] Brandt, "The Great Blizzard of '88," pp. 43, 94–95; Werstein, *The Blizzard of '88*, pp. 83–93.

[32] *New York Times*, Mar. 14 and 15, 1888; Waring, *Street Cleaning*, pp. 95, 97–98.

triple-team plows in opening its lines on the second day. Brooklyn and the other New York suburbs were as tightly blockaded as Manhattan on the 12th, but when the storm slackened, however, they were better able to resort to sleighs and horse sleds. With the re-opening of the Brooklyn Bridge on the 13th, the flood of traffic resumed. Albany, New Haven, Hartford, and Boston each suffered blockaded streets and railroads, as well as downed telegraph and telephone lines, yet none felt the shock as acutely as New York. Albany's 46.7 inches and New Haven's 44.7 inches did establish new (and much deeper) snowstorm records there, but that was not the major consideration. None of these cities had acquired the press of business centered in New York, and none had constructed elevateds to override the storms; moreover, none had provided a battery of city-operated horse carts and plows to open surface lanes. To see these costly safeguards suddenly frustrated was shocking, almost disillusioning, to self-confident New Yorkers. The editor of the *Times* was not alone in insisting on the speedy burial of all electric wires and in calling again for the early construction of a subway.[33]

The March snows melted rapidly, and Coleman kept his shovelers busy clearing the drains in order to prevent flooding. With the winter safely past, New York again relaxed. Along with the rest of the country, it enjoyed a relatively mild and dry winter in 1888–89. The only blustery storm came in January and the two-inch snow quickly turned to slush. The wind, however, again disrupted the dense festoons of wires overhead and prompted the city to launch the construction of underground conduits that summer. Despite a warm day on March 12, many New Yorkers reminisced over their experiences the year before,

---

[33] Brandt, "The Great Blizzard of '88," pp. 8, 43, 48, 77, 82, 94, 96; *New York Times*, Mar. 1888; *Weatherwise*, Dec. 1958, pp. 187–90, 211.

and the Brooklyn Academy of Photography assembled an exhibit of photographs of the Great Blizzard the next day. But it was a year later, after another relatively mild winter, that a threatening storm on March 4 stirred fears of a renewed blizzard. Farmer Dunn refused however to sound an alarm and the storm quickly fizzled out. Yet with earlier snows that year, in several other cities as well as New York, it raised a serious question as to the safe operation of the newly-introduced electric trolleys during snowstorms—a question only the future could answer.[34]

[34] *New York Times*, Mar. 15, 1888; Jan. 21 and 22, 1889; Mar. 3, 4 and 7, 1890.

A MOTORIZED PLOW, LOADER AND DUMP TRUCK TACKLES A DRIFT
(Courtesy of *American City*, 1923)

When snow blockades became costly hazards in thriving cities, their councils approved the purchase of machines to clear the principal streets.

CHAPTER FOUR

FROM SNOW-PLOWING TO SNOW REMOVAL

1890–1925

As the expanding cities switched in the early nineties from horse cars to electric trolleys and from messenger boys to telephones, urban residents in the northern states encountered new wintertime hazards. The increased volume of trade made all interruptions costly, and when a heavy snow not only stalled traffic but also felled wires and disrupted communications, small as well as great cities discovered the need for forthright actions. A few cities that suffered frequent and heavy storms responded early, but none matched the efforts made in New York to keep the streets open and traffic moving. Earlier storms had prompted the metropolis to take the lead in carting as well as plowing snow from the car tracks and had speeded the construction of elevateds to surmount the drifts; with the electrification of the horse car lines in the 1890s and the arrival of automotive traffic a decade later, New York had to tackle the job of clearing entire streets in the congested districts where even moderate storms threatened a blockade. Other cities, expanding with the electrification of their car lines, enjoyed an increased press of business and faced the task of snow removal in their commercial districts, where some also buried their electric wires. Goaded by heavy snows, Rochester and Buffalo adopted plans of action in the early 1900s, and a decade later, when a widespread blizzard battered several cities on the southern fringe of the snow country, Philadelphia summoned delegates from its fellow sufferers to plot new methods for snow-fighting. Few of the programs recommended in 1914 were new, but already new problems were emerging as the flood of automobiles increased; within another decade many cities would be tackling increased snow hazards with motorized forces.

Early Snow-Fighting Programs

Memories of the 1888 blizzard made New Yorkers apprehensive for many winters when snowstorms threatened. Street Commissioner James Brennan, who succeeded Coleman, devised a plan in 1892 for more effective snow removal along Broadway and other business streets. He assigned a number of blocks to each of his street-cleaning divisions, ordering them to suspend operations in their districts during a storm and to concentrate on the important task of opening the assigned arteries. As a result of the more efficient operations, he was able, when a storm dropped seventeen inches on New York in February 18 and 19, 1893, to boost the total number of cart loads of snow removed to a new high of 86,313 for the season and to keep traffic moving in the horsecar and trolley lanes and along Broadway. He hired gangs of immigrant Italians as shovelers to assist the cartmen, entrusting their *padrones* to distribute and retrieve the shovels and to deliver the wages at $.50 for eight hours, day or night. A group of residents along Fifth Avenue, critical of the abandonment of that avenue to sleighs, hired shovelers to clear the crosswalks and curbside drains at the 57th Street intersection, piling the snow in the center of the side street to assure a quick run-off. Pedestrians had rights, too, they maintained, and after an inspection, Brennan agreed to apply their approach to a longer stretch of the avenue in the next storm[1]

But New York's tremendous growth had not only brought increased pressure for uninterrupted trade along its principal arteries; it had also brought increased densities in poor immigrant districts on the lower East Side where a suspension of garbage collection for a day or two, while the carts removed snow from the avenues, left hundreds of thousands wallowing through disease-breeding mounds of filth. Alerted by Jacob Riis to the plight of the slums and the hazards they posed, the New York Academy of Medicine protested the diversions of street-cleaning forces from their principal job. Organized labor, fearing the competition of immigrant hordes in the depression year, got a law passed restricting city employment to full-fledged American citizens. A lighter fall the next winter enabled Brennan to meet these criticisms without sacrificing the snow-removal task, but 1895 brought a new administration and a new response to the total problem.[2]

As Police Commissioner, Theodore Roosevelt persuaded George E. Waring, a leading civil engineer, to accept the post of Commissioner of Sanitation and Street Cleaning. Waring, profiting from observations of practices abroad, not only bought improved garbage carts and, to enhance morale, uniformed his employees in white, but also charged the superintendents of his twelve districts with responsibility of collecting garbage in their areas throughout the year. He negotiated a new agreement with the horse car and trolley companies whereby each undertook to plow its tracks throughout and to remove snow from the entire street in certain districts. To escape the requirement of hiring only American citizens for the emergency tasks in the city's districts, Waring contracted

---

[1] *New York Times*, Jan. 10, 17 and 19, 1893.
[2] *New York Times*, Feb. 18, 1893; Waring, *Street Cleaning*, pp. 11, 100–01, 117–28.

with an independent company to supply all needed teams, carts, plows, and shovelers to remove the snow from certain areas. As a result, he practically tripled the number of loads removed in the first full year of operation, and more than doubled it again in 1896–97 when the mileage cleared was increased from the 22.8 of 1893 to 144.4. Contractors dumped more than 700,000 cartloads, and spent $445,000 in its removal; the city recorded only a moderately high total of 39.1 inches that winter, but the condition of the streets elicited praise.[3]

No other city in America, or Europe, could rival that feat. When a later snowstorm blanketed Washington, a "borderland" city, on March 4, 1893—the sixth snowy Inauguration Day in the nation's history—it cut the crowd of spectators to 10,000 shivering admirers of Grover Cleveland, but prompted no snow removal efforts. Memphis, another "border" city, astonished by a record-establishing snow of eighteen inches on March 16 and 17, 1892, had accepted it as an exciting historic landmark. All railroad, street, and river traffic came to a halt, and drivers kept their horses in their barns for two days, but members of the Ancient Order of Hibernians would boast for many a year that they had turned out and paraded through the drifts on Main Street that St. Patrick's Day. Of course Memphis, and Washington as well, were too far south to take snow seriously, but several cities in the snow country were now tackling the problem more aggressively in their business districts.[4]

The blizzard of February 1893 had been widespread; moreover, it was the first major storm to test the capacity of the newly electrified transit companies in several cities. In Pittsburgh, for example, with a ten-inch fall, a stiff wind piled up great drifts, driving most residents from the

[3] Waring, *Street Cleaning*, pp. 91–106.
[4] Hughes, *American Weather Stories*, pp. 89–114; *Memphis Evening Appeal*, Mar. 17, 1931.

streets; "the electric lines had a sorry time of it," The *Post Gazette* reported. Two trolley companies had acquired electric powered rotary sweepers to replace the horse plows previously in use; by sending them repeatedly along their tracks, they managed to keep most routes open. But on the central traction lines, operated by cables, the gripmen had difficulty in handling the levers on the steeper grades and had to suspend operations for a time. In Buffalo, blanketed with nearly three feet of snow, the trolley company battled "all night with mammoth snowplows" to keep a few lines open. As the plows, drawn by six horses (which fortunately had not yet been sold), piled up great ridges along their routes, the city hired shovelers to open and clear the crosswalks in the commercial district.[5]

Having disposed of most of its horses, the trolley company in Rochester had managed to keep only a few lines open during a succession of heavy snows in late December and early January of 1892–93. When the February blizzard hit, the company again brought out its reduced contingent of horse plows, but the accumulation of mounds in the central district threatened a blockade and forced the city to spend $15,700 to cart snow to the river banks. When the storms resumed two years later, however, the company was ready with an electric motor plow equipped with rotary blades to break through the drifts. The new plow, invented by Captain George W. Ruggles, a steamboat operator on Lake Ontario, proved its true utility in drifted sections and the company was able to maintain service on all routes. Encouraged by increased patronage, the company gave the city $2,000 for its assistance in removing snow mounds from the central district. Rochester

---

[5] *New York Times*, Feb. 20, 1893; *Memphis Evening Appeal*, Mar. 17, 1931; *Pittsburgh Post Gazette*, Feb. 20, 1893; *Buffalo Courier*, Feb. 20, 1893.

was thus the first to follow New York in this kind of collaboration.[6]

Buffalo, Rochester, Syracuse, and Oswego were becoming increasingly aware of their shared plight as urban centers in a region subject to heavy snows. The gusto with which some early settlers had boasted that current storms "had beat Vermont all hollow" had passed as newcomers from other regions and abroad outnumbered the old Yankees. Many had been attracted by the lush vegetation the Great Lakes helped to nurture, and a rich fruit belt had developed under the protection of Lake Ontario's stabilizing temperatures. But if the vast expanse and great depth of that lake forestalled the formation of an ice covering, the open and relatively warm water fed cold winds from the north with moisture and built heavy clouds that turned to snow as they moved over the colder land. The southern shore of Lake Ontario (and for slightly different reasons) that on the southeastern border of Lake Erie were snowbelts that now gained recognition, as regional weather records were accumulated and analyzed, far surpassing the snowfalls, in depth and frequency, of any other densely settled region in the nation.[7]

Thus it was not surprising that while most cities in the broader snow country continued to rely on the sun to remove their snow mounds, several in the New York snowbelts of the early 1900s began to require their trolley companies to plow and distribute the snow so as not to obstruct other traffic. New "V" plows, with shields or flanges designed to push the snow farther aside, had won popularity on some steam railroads, and adaptations had

[6] *Rochester Herald*, Dec. 26, 1892; Jan. 11, Feb. 20, 1893; Jan. 31, Mar. 27, 1894; Rochester Executive Board, *Report* (1893), p. 141; Wm. R. Gordon, *Ninety-four Years of Rochester Railroads* (Rochester: n.p., 1975), pp. 32–44.

[7] B. L. Wiggin, "Great Snows of the Great Lakes," *Weatherwise*, Dec. 1950, pp. 125–26; See also the Narrative Climatological notes attached to the *Local Climatological Summaries*, Rochester, 1900–1951.

appeared among horse plows; unfortunately their operation in deep snows required double or triple teams, and as trolleys displaced horse cars, the supply of horses was diminishing. The railroads, particularly in the West, had meanwhile acquired powerful rotary plows to break the drifts that sometimes clogged mountain passes or valleys in the great plains. Lively descriptions of their operation during a blizzard in the Northwest in 1897, published in *Cosmopolitan* and other journals, alerted readers in snow-belt cities to the technological improvements.[8]

But attaching plows to the bobtail trolleys then in use seemed hazardous. Attempts to design a plow that could be pushed around sharp street corners without derailing the car were not mastered for another decade. Rotary sweepers were less hazardous but of little use in deep or wet snow, as Pittsburgh had discovered. The editor of *Scientific American* was pleased to report, late in 1905, that trolley companies in Boston and Worcester had introduced a few heavy trolley cars equipped at both ends with new Wilder Radial plows, and had demonstrated their usefulness in battling snowdrifts that January.[9]

Responsible municipal as well as transit officials were eagerly looking for technological breakthroughs. Commissioner Waring of New York had tried two different movable "snow melters" in an effort to dispose of the snow without carting it to the river piers. Although neither proved practical, he urged continued search for such a solution. When the elevateds substituted electric motors for their steam engines, only to find the live third rail frequently coated by ice in a sleet storm, a "sleet cutter" designed to maintain contact without producing terrifying flashes, solved the problem. But the great achievement in New York was the opening in 1904 of the first stretch of its

---

[8] *Cosmopolitan*, 23 (1897): pp. 363–67; *Scientific American*, 64 (1891): p. 130; 87 (1902): p. 641.

[9] *Scientific American*, Supplement, Dec. 9, 1905, p. 25033.

long-projected subway system. The secure haven and unimpeded service it supplied during the first severe storm of January 1905 prompted many to agree with the *Times* reporter who declared the "The Subway proved to be worth about all it cost," which was quite a sum.[10]

Though less determined in snow-fighting, other cities were taking actions that brought some relief. Boston had completed a mile-long subway under its central district five years before and, by routing several trolley lines into that tunnel or onto six miles of elevated track, had greatly relieved traffic congestion in winter as well as summer; and following New York's example, it had buried telephone and electric wires in underground conduits along some principal streets. Boston's extensive trolley system, equipped with large cars able to push snow-plows, had eliminated the need for many horse-plows and kept traffic moving even in the face of occasionally heavy snowstorms.[11] Chicago had opened its South Side elevated line in time for the Fair in 1893 and had the West Side and North Side lines and the "Loop" in operation four years later. This year-round improvement greatly relieved the blockades experienced by its extensive cable-car system during heavy snowstorms when ice or packed snow, clogging the slots, prevented the cars from securing a sure grip on the underground cables. Chicago had vied with New York, Boston, and several other snow-plagued cities in speeding up the removal of overhead telephone and electric wires into underground conduits, chiefly for aesthetic reasons in preparation for its Fair but also for safety in snowstorms. Prominent among the cities where the electric companies, sometimes in collaboration with the municipality, had made early efforts to bury their wires in underground conduits were Buffalo and Rochester in the snowbelt, Philadelphia and Washington on the southern fringe of

[10] Waring, *Street Cleaning*, pp. 106–09; *New York Times*, Jan. 5, 1905.
[11] *New York Times*, Feb. 2 and 3, 1898; Mar. 2, 1905; Jan. 14, 1913.

From Snow-Plowing to Snow Removal

the snow country, and Detroit and Milwaukee in the Midwest. Although Chicago, now the nation's Second City, annually appropriated $25,000 for snow removal in the early 1900s, when a snowy January in 1910 exhausted the fund, the council discontinued the service and relied on its elevateds and its steam lines for necessary transit.[12] Philadelphia continued to depend more heavily than other leading cities on its steam commuter trains, and the municipality generally contented itself, as did Washington and many lesser cities, with supplying horse-plows to clear sidewalks along its principal streets after a storm. A clarification of earlier court decisions fixing the responsibility on cities for safety on pavements and crosswalks speeded this development.[13]

Municipal Responsibilities Assumed

The three major snowbelt cities could not escape so easily. When Rochester suffered a record-breaking fall of thirty-six inches in as many hours on March 1 and 2 in 1900, even the experienced snow-fighters of its trolley company, equipped now with an improved Ruggles plow and a rotary sweeper, were temporarily stymied. As the depth reached forty-three and a half inches the next day, businessmen who ventured out relaxed by tossing snowballs at their friends. Rochester could not, however, rely exclusively on "that old fellow Boreas" and mustered an army of shovelers and a fleet of bobsleds to remove the snow from the central blocks of Main and State Streets. The city spent only $10,000 for snow removal that year because of the late date of the storms, but two years later, when a succession of

---

[12] *Municipal Journal*, Feb. 2, 1910, p. 181.
[13] *Municipal Journal*, Jan. 25, 1911, p. 133; Jan. 4, 1912, p. 31; Mar. 14, 1912, p. 386; John Simpson, "Municipal Liability for Snow on Crosswalks," *Public Works*, Feb. 1924, pp. 62–64.

storms started in January and continued intermittently for two months, it tripled its expenditure and dumped more than 70,000 cubic yards of snow into the Genesee River. The city engineer marshalled a force of seventy-five teams and carts and three hundred shovelers to battle a snowstorm the next January and added four new streets to the snow-removal schedule because of the wider demand for that service.[14]

Rochester's deliberate action in formulating a snow-fighting plan in 1904 (second only to New York's much earlier plans) reflected several congruent circumstances. As one of the principal snowbelt cities, it had a recurrent problem, and in 1904 elected the third and ablest of three "good government" mayors, James G. Cutler. Cutler had scarcely been sworn in before the outbreak of one of the city's most disastrous fires in the central business district, in the middle of a snowstorm on February 26, demonstrated a pressing need for leadership. Not only was the fire-fighting equipment found to be antiquated, but it was hampered by the chaotic furrows and hillocks into which the snow had been plowed on crucial downtown streets. The trolley company had recently been acquired by a Philadelphia trust, which complacently relied on the sun to clear its tracks of snow. The company's inaction during a moderate storm in January had prompted a sharp warning from Mayor Cutler that its Rochester franchise and the Rochester weather both required prompt action when a snowstorm hit. As a result, the inexperienced Philadelphians sent out plows when the February storm started, but their major effect was to block some vital streets and to obstruct the fire engines. A plan to coordinate the snow-plowing and snow-removal functions of the city and the trolley company was needed, and Cutler ordered City Engineer E. A. Fisher to prepare a blueprint. He further

[14] *Rochester Democrat & Chronicle*, Mar. 1, 2, 3 and 4, 1900; Gordon, *Ninety-four Years*, pp. 42–44.

directed him to extend the sidewalk-plowing program from 12 to 83 streets, a service unrivalled elsewhere.[15]

None of the other snowbelt cities took such forthright action. Syracuse, which received heavy snows during most of these years, was content to clear the central crosswalks and a few downtown sidewalks, relying on its trolley company to plow its tracks. Oswego, Utica, and Albany still accepted storms as inevitable interludes to be patiently endured.[16] But Buffalo, the busiest and the hardest hit, especially in 1910, was finally forced to shoulder the task of hauling snow in order to resume activities in its business district. Its expenditure of $15,000 in 1909, chiefly in clearing crosswalks and removing mounds from downtown intersections, had to be doubled the next year in the face of heavy storms in January and February.[17]

The winter of 1908–09 had deposited heavy snows on widely scattered cities throughout the snow country; it set only one record but produced some dramatic responses. Minneapolis and St. Paul together spent over $50,000 combatting snowfalls that season; Chicago, caught in the same wintry onslaught, topped that sum for the first time in its history. Both Detroit and Pittsburgh received heavier than usual blows and used dump carts to help clear their business streets. Denver received a record-breaking 118.7 inches in a long old-fashioned winter that many of its residents enjoyed. Far to the east, Boston, less tolerant of interruptions, re-doubled its efforts and spent a total of $114,026 on snow-fighting that winter. In Washington, a contingent of

---

[15] *Rochester Democrat & Chronicle*, Jan. 8 and 10, 1905; Jan. 9, 1906; *Proceedings of the Rochester Common Council* 1903, p. 19 and 1905, p. 18; *Municipal Research*, (Dec. 1936) p. 45; McKelvey, "Snowstorms & Snow Fighting" *Rochester History*, vol. 27 (Jan. 1965): No. 1, pp. 9–11.

[16] *Municipal Journal*, Jan. 25, 1911, p. 133; Feb. 22, 1911, p. 266; *Local Climatological Summaries* for Buffalo, Rochester, and Syracuse, 1900–1950; *Municipal Reports for the City of Syracuse*, 1910, 1912, p. 54. Its expenditures in 1912 reached $15,568.

[17] *Buffalo Commercial Advertiser*, Feb. 1910.

federal employees helped city street workers in removing a ten-inch snow from Pennsylvania Avenue in front of the White House to clear the way for the inaugural ceremonies for President Taft on March 4, 1909.[18]

But no city faced the problem as soberly as New York. Commissioner John M. Woodbury, who succeeded Waring, maintained daily contact with the local office of the U.S. Weather Bureau (finally established on a civilian basis in 1891 within the Department of Agriculture to assume the climatological functions of the Army Signal Service). Even a three-inch fall produced a vigorous response there, and Woodbury reported spending $309,000 on three storms early in January 1904 that dropped a total of only 10.3 inches, but resulted in the removal of 1,566,350,741 cart loads from 173 miles of the city's streets. Three years later, the cost for the season had reached nearly two and a half million dollars. The department put its road scrapers into service the next winter, driving them in pairs down one side and back the other to scrape the entire street surface, while shovelers kept the drains open. An experimental attempt to dump the snow into the trunk sewers through enlarged manholes equipped with a heating device to speed runoff proved disappointing, as the accumulated dirt threatened to clog the pipes. New dumping stations along the river had to be found.[19]

Several cities on the periphery of the snow country assumed new responsibilities and some tried new experiments. A resident of Elmira, New York, attached a plow in front of his Overland runabout in January 1910 and cleared the street of snow in his block to demonstrate the feasibility of motor-driven plows. Cincinnati's street-cleaning depart-

---

[18] U.S. Census, *General Statistics of Cities for 1909*, (Washington, 1913), p. 135; *Weather Record Book*, pp. 74–77; Hughes, *A Century of Weather Service*, pp. 93–110.

[19] *New York Times*, Dec. 15, 1902; Jan. 30, 1904; *Municipal Journal*, Apr. 21, 1909; Feb. 2, 1910; Whitnah, *History of U.S. Weather Bureau*, pp. 58–61.

ment spent $23,432 in carting snow from its downtown district hit by four rare storms in January and February of that year. Kansas City was caught unprepared for the record-shattering twenty-five inches of snow that blocked its streets on March 23–24, 1912. Some parts of the city were not so deeply blanketed, however, and the trolley company, by running cars equipped with rotary sweepers and assisted by fifteen hundred shovelers in two successive shifts, kept a few lines in operation. With seven hundred men and two hundred fifty teams, the city tackled the job of opening the principal business streets; it even pressed two recently acquired asphalt dump trucks into the snow-removal job. But in the stiff wind, loading a dump car, as one reporter observed, was "like shoveling fleas." As the depth increased in some sections, motor trucks had to engage the help of double and triple teams to make essential deliveries. The deluge "equaled Maine's best," the *Kansas City Times* wryly declared. Fortunately the temperature was rising, and by opening the water hydrants in key districts on the third day, the city finally brought its harshest winter to a close.[20]

Denver, more experienced with heavy snows, was ready with street-grading machines to clear the drains and gutters after a storm in January 1913, and marshalled a battery of dump carts and shovelers to help remove the worst mounds. That, however, was only a foretaste of the storms it received early the next December when a fall of 45.7 inches in six days buried the city under a record-breaking blanket of snow that halted all traffic. The tram company had got its rotary sweepers out shortly after the snow began to fall on Monday and kept its routes open until Thursday when a downpour of snow caught most of the cars in the streets. The sudden avalanche, which dumped nearly four feet of snow in two days, left the city on Saturday "completely in the grip of the snow," as the

---

[20] *Municipal Journal*, Mar. 2, 1910, p. 338; Mar. 16, 1910, p. 378; *Kansas City Times*, Mar. 25, 1912; *Kansas City Star*, Mar. 24 and 25, 1912.

official report declared. The city improvement department jointed the street-cleaners and the tram company, mustering three hundred fifty carts and six hundred men in a Herculean effort to break the drifts and clear the streets, a task that continued for another ten days.[21]

Cleveland's ordeal in the record-breaking storm that dropped twelve to eighteen inches on the city and its suburbs in November 9–11, 1913, highlighted the plight of many cities on the fringe of the snowbelt. Accustomed to battling moderate snows with trolley plows, the Cleveland Railway Company managed to keep several lines operating until the raging gale toppled several hundred giant telephone and electric poles across its routes, halting all traffic on many streets. Milk, produce, and coal deliveries, maintained with great effort after the first onslaught, were stopped on the second day as horse carts and heavy trucks became stalled in the drifts. Fear of a milk famine prompted one company to send men on horseback to its suburban dairy with orders to bring milk back by sleds to a point where motor trucks could pick up the cans for delivery to the plant and a few hospitals. Passengers in marooned trolleys and interurbans had to be rescued with wagons and sleds drawn by double teams. Coal and meat wholesalers used double teams to haul emergency supplies to hotels and other distribution centers. "If Clevelanders expect to eat in the next few days," *The Plain Dealer* warned, "they will be obliged to walk to the stores and carry food home in market baskets." The problems were not that simple, however, for most of the stores that managed to open on the second day saw their supplies quickly depleted. The danger of electrocution from fallen wires was miraculously reduced as utility crews untangled or removed the wires under police guard, and only one death was reported by the press. Accidents and fatalities, particularly among

---

[21] *Municipal Journal*, Feb. 13, 1913, p. 238; March 19, 1914, p. 357; Albert W. Cook, "Snowfall at Denver," *Weatherwise*, Oct. 1952, pp. 104–05.

the men manning boats on the lakes, were numerous, however, and the physical damages to utilities and merchants were the heaviest ever experienced in Cleveland. Only the fortunate arrival of warm weather on the fourth day brought a promise of relief.[22]

Winter Sports Revived

The havoc wrought by Cleveland's great blizzard was too traumatic to permit a mention by the distraught editors of its *Plain Dealer* of any recreational sidelights. Yet newspaper accounts of heavy snowstorms in cities more accustomed to such ordeals often tried to relieve the gloom by taking note of some light-hearted citizen responses. The sound of sleighbells on track-free streets or in the parks was becoming less noticeable in New York City as motor traffic increased after the turn of the century, but it still inspired nostalgic comments in some snowstorm editions.

In frigid but dry winters skaters continued to congregate in great numbers on ponds and other urban widewaters, and ice-yacht regattas re-appeared on the Hudson and other natural rinks near New York and Boston in recurrent arctic seasons. A succession of cold dry winters at Rochester saw the formation in 1899 of a club by a dozen ice yachtsmen on Irondequoit Bay; not to be outdone, a score of Syracuse sportsmen formed a similar club on Onondaga Lake a year later. Ice hockey became popular among skaters, and rival athletic clubs from Boston and Montreal among others played inter-city contests. The number of such events listed in the index of the *New York Times* increased dramatically in these years.[23]

[22] *Cleveland Plain Dealer*, Extra, Nov. 11, 1913; Dana T. Bowen, *Shipwrecks of the Lakes* (Daytona Beach: n.p., 1952), p. 354.
[23] *New York Times*, Mar. 7, 1896; Jan. 14, 1900; *Rochester Democrat & Chronicle*, February 19, 1900; *Syracuse Post Standard*, Sept. 23, 1901; *New York Times Index*, 1905–1912.

Thus the leader of a newly formed league of hockey teams at Rochester, inspired by a visit to one of Quebec's annual winter carnivals, dating from 1894, persuaded the city to stage an ice carnival in its Genesee Valley Park in February 1910. Several weeks of sub-freezing weather and moderate snows had enabled its wintertime enthusiasts to erect an imposing ice tower thirty feet high surrounded by a number of igloos and castellated enclosures. With the Rochester park band playing for skaters on the glassy surface of a natural rink under a canopy of Japanese lanterns hanging from electric wires, thousands of its citizens celebrated the newly proclaimed rites of winter until a mild thaw set in towards the end of the month[24] Other Rochesterians were excited that February by an "endurance run" promoted by the automobile club, which sent twenty-two hardy motorists on an excursion to Syracuse and back in an effort to prove that it was not necessary to store and jack up one's car in winter months. All entrants successfully completed the two-day ordeal although nine had to be hauled out of drifts by bantering farmers. Both the skaters and the motorists suspended activities when March brought a record-breaking snow deposit totaling forty-two inches.[25]

The delight many city-dwellers still took in a blustery snowstorm sometimes hampered effective action. A sudden deposit of 11.8 inches on New York City on December 24–25, 1912, gave it another White Christmas, but many car drivers who came in to enjoy the excitement became marooned in the snow—the first public mention of that serious new snowstorm problem—and complicated the task of the street-cleaners who were attempting to keep the avenues open. That department was further handicapped

---

[24] *Rochester Democrat & Chronicle*, Feb. 19, 1910; *Rochester Herald*, Feb. 6 and 20, 1910.

[25] *Rochester Herald*, Feb. 2, 3 and 4, 1910; *Local Climatological Summary*, (1951).

From Snow-Plowing to Snow Removal

as men hired to load its carts deserted to get free Christmas dinners at various charities, and only nine hundred of its two thousand dump carts were in use that day. "Old Santa would have been at home in New York yesterday," as a *Times* reporter noted, but his motorized successor had difficulty making late deliveries and many accidents marred the season. The city had to recruit 4,500 additional men on the day after Christmas and supply pickaxes as well as shovels to clear the streets.[26]

Inter-city Trends and the First Snow Conference

Big storms continued to provide the chief goad to urban wintertime programs, but new technological developments both in transportation and communication were influencing the character of that response. The increased use of motor cars in winter months created new hazards in the streets and gave rise to new demands for municipal efforts—as distinguished from transit company action—to keep traffic moving. Fortunately, the newly-introduced motor dump trucks, as Kansas City demonstrated in 1913, supplied a new aid in snow removal, and a year later, when a rare fifteen-inch snowfall blanketed St. Louis, its street-cleaning forces not only followed its neighbor's example in the use of dump trucks, but also pressed several newly acquired motor-driven road scrapers into use and cleared the streets in three days.[27]

Because of the wide dispersal of municipalities and the localized character and traditions of their administrations, news of these and other innovative experiments would have spread more slowly had it not been for the develop-

---

[26] *New York Times*, Dec. 25, 1912, pp. 2–3; Dec. 26, 1912, pp. 1–14; Dec. 31, 1912, pp. 4–6.

[27] *Municipal Journal*, Mar. 12, 1914, pp. 357–58; Dec. 10 and 19, 1914, p. 829. *Weather Record Book*, pp. 75–76.

ment of a number of specialized service journals in the early 1900s. The newly launched *Municipal Journal* began to include brief reports from various cities describing their programs. One article listed the names of seventeen plows already on the market and described one which had demonstrated its ability in New York to do the work of twenty-five shovelers. The *Journal's* editor observed in November 1913 that, while most northern cities had ordinances requiring residents to clear snow from their sidewalks and several engaged horse plows to do this work, only a few regularly attempted to cart it from the streets, and none approached New York in that effort. Other journals took note of the problem. *Scientific American* had reported technological improvements in snow plows, sweepers, and melters in the nineties and now became interested in the organizational strategy for snow-fighting. It also published numerous articles describing improvements in the insulation and installation of underground wires. The *American City* featured reports of the practical use of snow-plows and other equipment advertised in its pages. In addition, enterprising firms venturing to produce motorized and detachable plows, dump trucks, and other snow-fighting equipment, provided a significant promotional backing for the more aggressive and widespread municipal response to snowstorm hazards.[28]

Of course the storms and the problems were never the same. New York, visited again by a succession of storms in January and February 1914, found its forces handicapped, in a city already converting to motorized traffic, by a shortage of teams and carts. Its officials saw new storms arrive before earlier deposits could be removed and had to spend $2,500,000 to clear the streets. Philadelphia suffered

[28] *Municipal Journal*, Dec. 7, 1910, pp. 769–73; Nov. 6, 1913, pp. 640–41, Mar. 12, 1914, pp. 335, 365; *Scientific American, Supplement*, Sep. 25, 1891, p. 13118; Mar. 14, 1914, p. 281; *American City* XII (1914): p. 342.

From Snow-Plowing to Snow Removal

less but did not escape the widespread storms of 1914. It had received a record breaking twenty-one inches in December 1909, and the threat of a repetition was alarming. Teams and carts were becoming less plentiful there, too, and the old practice of hiring carts and shovelers as needed when a storm hit proved inadequate.[29]

Recognizing the new snow hazards to motorized traffic, Morris L. Cooke, director of public works in Philadelphia, took the lead in staging a convention of public works officials at its monumental City Hall in April 1914 to consider urban snow-removal problems. A score of concerned officials attended this pioneer gathering, among them J. W. Paxton from Washington who served as chairman. J. T. Fetherston, now in charge of snow removal in New York, stressed the need for advanced planning, while others with experience in combatting snow at Boston, Baltimore, Pittsburgh, and Scranton reported on their programs. For some reason Buffalo and Rochester, along with New York City, the richest in experience, failed to send delegates, but Cooke had assembled reports on practices abroad, and the conference was informative.[30]

All agreed, as Paxton reported, that the first essentials were advanced planning and effective organization. These called for agreements between city departments concerning their respective contributions of equipment, supervisors and laborers, if the city planned to do the job, or, if it was to be let out, firm contracts with responsible agents. In large cities a division into zones or districts was recommended, with specific assignments distributed in advance. New York's recent experience, as well as that of Pittsburgh, emphasized the merits of making a prompt start

---

[29] *Municipal Journal*, Dec. 10, 1914, pp. 827–29; *American City*, 11, (1915): pp. 280–84; Conrad P. Mock, "Philadelphia's Greatest Snowstorm-Dec. 25–26, 1909," *Weatherwise*, (Dec. 1951): pp. 134–35.

[30] *New York Times*, Feb. 14, 1914; *American City*, 10 (1915), pp. 494–95; *Scientific American, Supplement*, Mar. 6, 1915, pp. 159–60.

"as soon as snow had covered the pavements" rather than trying to economize by waiting for the end of major storms. The recommendations further urged: (a) careful consideration of the capacity of the sewer system to carry a large run-off and to handle bulk dumping of snow; (b) adoption of a flexible budget to avoid fiscal delays in a crisis; (c) negotiations of firm agreements with the transit companies concerning the plowing and removal of snow from their tracks. E. D. Very of New York stressed the need, in determining which streets should be cleared, to consider sanitary precautions as well as traffic flow. A representative of the Pennsylvania Railroad cited its successful use of a movable snow melter in clearing snow from clogged freight yards and passes. Another delegate called for a reconsideration of the use of salt, as in Liverpool, London, and Paris, to speed snow removal.[31]

Publicity on that first conference spurred other suggestions. P. A. Hutchison of New York proposed the use of power shovels to remove snow after it was plowed into furrows, and the substitution of motor trucks or "flatcars" for the inefficient horse carts. He urged the development of a suction plow to carry the snow back to a dump car as on some steam railroads. A Philadelphia manufacturer quickly announced that such a plow had already been produced and that several street commissioners had placed orders, but that New York seemed to think the first need was for more teams and carts.[32] The choice depended, as Fetherston noted, not only on the size of the storm, but on the character of the job to be performed. In New York, for keeping all streets clear even of slush for sanitary reasons, an army of shovelers and many carts were more desirable than power shovels and suction or rotary plows. Fetherston's policy was to tackle each storm as it started and to increase his force as it developed in order to avoid major

---

[31] *Scientific American Supplement*, Mar. 6, 1915, pp. 150–51.
[32] *Scientific American*, Mar. 14, 1914, p. 281; Apr. 4, 1914, p. 283.

pile-ups that might require heavier equipment. For almost six years the weatherman would cooperate.[33]

Several cities adopted and implemented a snow-fighting strategy. Boston determined to rely on the one thousand-man force and three hundred carts in its street-cleaning department to handle snowfalls up to six or eight inches; for heavier storms it authorized each of six district superintendents to hire additional workers as needed. Philadelphia designated twenty miles of streets where snow-removal operations should be pressed and assigned special tasks to two hundred of its regular employees who, as a storm developed, would hire additional workers and direct shifts in one or another of the city's nineteen districts, each with its approved snow dump. Chicago organized its street-cleaning forces for snow removal with plows and carts first used in 1912.[34]

Two widespread storms in February and March 1916 brought new adjustments in each of these and several other cities. Syracuse had received a heavy deposit in December, and reports of its efforts to dig out may have helped alert the new superintendent in New York to add a number of the newly-advertised motor plows to the department's arsenal. The efficiency with which the first contingent of plows piled the snow of a January storm into windrows to be shoveled into carts prompted C. L. Edholm to order additional motor-plows until he had a fleet of ninety-six by the end of the season. With the assistance of some 14,700 workers he cleared a total of 946 miles of streets in a winter that brought snowfalls about 50 per cent above the average. Despite the increase he was able, he boasted, to keep the cost down to $2,521,000 for the season.[35]

---

[33] *Municipal Journal*, Dec. 10, 1914, pp. 827–29; June 10, 1915, pp. 805–06; *American City*, 11 (1915), pp. 280–84.

[34] *Municipal Journal*, Dec. 10, 1914, pp. 830–32; *Government of the City of Rochester*, (New York: Bureau of Municipal Research, 1915), pp. 397–407; *American City*, Aug. 1937, p. 49.

[35] *Scientific American*, Dec. 15, 1916, pp. 547, 560.

Other cities were likewise adopting motorized equipment. Boston acquired three motor plows in 1916 and used them effectively after a near-record storm in opening the major streets, which were more completely cleared later by the city's twenty horse-plows. The Mercury Manufacturing Company of Chicago, and the Good Roads Machinery Company in suburban Philadelphia filled orders from several cities, including Buffalo and Rochester, each of which battled snowfalls that more than doubled New York City's total for the season.[36] Milwaukee and Minneapolis as well as Philadelphia acquired motorplows, but it was Newark which apparently made the first use of its Keystone traction road building shovel to assist in clearing the streets after a ten-inch fall the next winter. Seattle, on the other hand, hit by a record-breaking snow of 32.8 inches on January 31–February 2, 1916, three times its annual average, made a valiant effort, with the aid of seven snow plows and eight hundred shovelers to keep traffic moving on the first day. But as the snow continued to fall, blockading the streets, collapsing many roofs, and finally crashing the dome of St. James' Cathedral, the astonished residents suspended normal activities for several days, and accepted the storm as an historical anomaly.[37]

Rival interests were battling for preference in several cities. Spokane, which seldom suffered heavy snows, joined the list of cities that regularly plowed their sidewalks, while Milwaukee spent $22,612 in a determined effort to clear its downtown street crossings in 1916. Chicago developed a plan that enabled it to mobilize its plows and carts within an hour by telephone and carted some 22,000 loads from the Loop district early in 1917, but the next winter's near-record

---

[36] *American City*, XIII (1915): pp. 371, 463; XVII (1917): p. 316; *Rochester Times-Union*, Jan. 19, 1917.

[37] *Municipal Journal*, 44 (Jan. 2, 1918): p. 31; "Seattle's Great Snowfall of 1916," *Weatherwise*, April 1962, p. 81; *Seattle Daily Times*, Feb. 1, 2, 3, 4, 5 and 6, 1915.

From Snow-Plowing to Snow Removal

deposit of sixty-four inches brought a paralyzing blockade. By a hasty recruitment of extra carts, sleds, and dump trucks, and by the employment of over three thousand shovelers, the city was able to remove 716,372 cubic yards of snow in three days, boosting the outlays for the year to an unprecedented $315,409. That struggle, chiefly with hand labor, prompted Chicago to order twenty-five plow blades to be installed on a new fleet of five-ton dump trucks to forestall similar disasters.[38] Rochester, enjoying a relatively open winter in 1917–1918, nevertheless doubled the number of loads removed and kept its streets open to the 15,000 motor vehicles now registered in the county. With two motor-plows, dump trucks and a sand cart, it was trying to meet the demands of an extremely vocal automobile club, but its snow-fighting plan also provided for the reservation of five streets leading out to the hinterland as "snow roads" for the convenience of hucksters from the country and sleigh riders from the city. Snow-plows clearing other streets had to raise their blades in crossing these routes, which provided good sleighing for almost four months in that frigid year. The city paid $120,928 to clear its other streets and sidewalks without complaint.[39]

Harsh Winters Develop New Hazards

The blizzards of 1917–18 had plastered Juneau, Alaska, with a record-breaking 246.3 inches, but they had directed their most unusual blows at cities in the mid-border states. Cairo, as well as Chicago in Illinois, Chattanooga as well as Memphis in Tennessee, had received record storms, but it

---

[38] *Municipal Journal* 42 (Jan. 11, 1917): pp. 49, 112; 44 (Jan. 12, 1918): p. 32; *American City*, Dec. 1922, pp. 493–95; Aug. 1937, pp. 48–49.

[39] *Report on the Problem of Snow Removal in City of Rochester*, (Rochester: Rochester Bureau of Municipal Research, Inc., 1917); *ibid.*, 1918.

was in Louisville, Kentucky that the twin blizzards of December 7–8, 1917, and January 14, 1918, would be longest remembered. The unprecedented deluge of fifteen inches in as many hours on Saturday, December 7, not only halted all incoming and outgoing trains, but blockaded most of the trolley lines as well. A force of 1,000 shovelers was increased to 1,500 on the second day and carted 2,000 loads of snow from the business district but re-opened traffic on only a third of the city's car lines. The schools and many shops were closed on Monday. The most threatening hazard, however, was from the shortage of gas and coal for heating use in the face of unexpectedly frigid temperatures. A warming trend relieved the emergency and allowed coal dealers to replenish their supplies before the second blizzard hit in January, but Louisville would not soon forget the winter of 1817–1918, the coldest as well as the snowiest in its long record, which also saw the Ohio River frozen over for several weeks.[40]

The entire Northeast suffered unprecedented hardships during that harsh winter of 1917–18. Although heavy snows fell only in the Ohio Valley, the cold wave spread eastward, plunging temperatures to record-breaking January lows from Cincinatti and Buffalo eastward to Pittsburgh and Washington as well as Rochester and Syracuse. For New York City and Boston and a half-dozen other cities, including Charleston, South Carolina, it was the coldest winter on record. Moreover, the suffering was compounded by a coal shortage that reached crisis proportions. The shortage was heightened by the wartime emergency that had glutted rail and port facilities, blocking the flow of military and food supplies to Europe. To break that vital bottleneck, Dr. Harry Garfield, President Wilson's Fuel Administrator, banned all industrial use of coal for five days in mid-January. By checking the flow of

[40] *Louisville Courier-Journal,* Dec. 9 and 10, 1917; Jan. 15, 1918; Feb. 1, 1966; *Weather Record Book,* pp. 54, 75.

manufactured goods, Wilson and his advisors hoped to speed the loading and unloading of essential cargo, and clear the tracks for coal and military supplies. The coal shortage had been mounting for several months, and dealers, objecting to a wartime ceiling of $2 a ton, had been slow to make deliveries. Household coal bins were empty or depleted; unable to shift to or find wood supplies, people were carrying emergency rations of coal home in sacks. Fortunately the week-long embargo of factories sufficiently relieved railroad congestion to permit a resumption of fuel and other deliveries by the 22nd and brought a relaxation of other restraints three weeks later.[41]

Another east-coast blizzard early in February 1920, and a succession of heavy snows inland in that and succeeding years, finally forced most northeastern cities to mechanize their snow-fighting forces. A fall of twenty-two inches in the course of two days, interspersed with crusts of sleet, dealt New York City a hard blow. Its recently-acquired motor trucks and plows "spun their wheels" in an effort to push the snow aside. The department brought out one of its steam shovels to clear a drifted street near a dumping place, but a renewed experiment with an asphalt burner as a snow melter again proved disappointing. The department, however, was now convinced of its need for new equipment and placed an order for one hundred five-ton Holt tractors mounted on movable treads like army tanks and able to push heavy piles of snow in winter as well as gravel or dirt in summer. Boston had received the main brunt of the 1920 storm, which dropped over three feet in most of its streets, breaking all records since 1893. The city's three motor-plows, assisted by a battery of horse-plows, made a valiant effort to keep some lanes open. But as the snow, interlayed with sleet, reached a depth of nineteen inches on the second

---

[41] James P. Johnson, "The Fuel Crisis, Largely Forgotten, of World War I," *Smithsonian* VII, (Dec. 1976): pp. 64–70; *Weather Record Book*, pp. 44–54.

morning, all traffic, even that of heavy trucks, came to a halt. As the storm continued, the tie-up was pronounced the worst suffered by the Bay City in two decades. To break the drifts, an engine pushed a locomotive crane into the North Station and loaded a string of flat cars with packed snow. Steam shovels seemed the only solution, and the city leased several from construction firms to help reopen its principal arteries. Boston, too, would upgrade its snow-fighting forces after the snow melted in March.[42]

The sheer weight of that storm alerted other cities to the need for more powerful plows and other mechanized equipment. Philadelphia and Chicago each ordered a number of caterpillar tractors and plow-blade attachments, and Chicago used one to operate the newly invented Barber-Green snow loader, which it tried out successfully in December 1920. New York and several other cities bought snow loaders that winter; the former achieved new efficiency when it cleared its streets of a 12½-inch fall on February 19 in twelve hours. That accomplishment combined several innovations. Alerted by the Weather Bureau, Chief John Kenlon of the Fire Department had sounded the new "14–14" snowstorm signal that called thousands of men in his department and the snow-removal forces to active duty. By its prompt dispatch of the new fleet of tractor-plows, snow loaders, and dump trucks, purchased at a cost of almost $2,000,000, the snow-removal department kept the main thoroughfares open and saved the city an estimated $10,000,000 in business. The close cooperation of the police and fire departments warded off threatened losses from gale-swept fires and gave the city a new sense of security.[43]

[42] *New York Times*, Feb. 6 and 7, 1920; *American City*, Oct. 1920, p. 360; Dec. 1920, pp. 583–84, 667; Oct. 1923, pp. 377–78; *Public Works*, Jan. 8, 1921, p. 37.

[43] *Scientific America*, Feb. 28, 1920, pp. 219–20, 232–33; Jan. 29, 1921, p. 85; Feb. 19, 1921, p. 153; *Public Works*, Feb. 19, 1921, pp. 165, 193; *American City*, April, 1921, p. 447; Oct. 1923, pp. 347–48; Nov. 1923, pp. 479–82.

From Snow-Plowing to Snow Removal

The rapid increase in the production of automobiles and trucks in the postwar years, with car registrations mounting to nine million by 1920 and to seventeen million by 1925, created a new demand for dry streets in winter months. The practice of storing cars on jacks in the stable had passed, and with it the faithful horse to pull a sleigh, and patience with horse-drawn plows as well. City after city ordered plow blades to be attached to street-cleaning or garbage trucks for extra duty in winter months. Paralyzed by a record-breaking storm that dropped a total of twenty-eight inches of snow in January 27–28, 1922, the old delight in Washington in sleighing was forgotten as residents shivered in stalled trolleys and visitors struggled to free their marooned automobiles. The collapse of the roof of the Knickerbocker Theater, crushing or smothering ninety-six of its occupants under the heavy weight of snow, dispelled complacency. Congress finally faced the necessity of enabling the city to acquire plows to clear its streets and take other wintertime precautions. Albany, Rome, and Utica in upstate New York, Hartford, Providence, and Worcester in New England, and Toronto in Canada joined the list of cities that had earlier assumed responsibility for clearing snow from business streets. In several of these towns, local Chambers of Commerce had to raise funds to buy the plows and snow loaders.[44]

A questionnaire circulated by the journal *Public Works* tabulated the equipment of one hundred cities. It reported that 53 already had snowplows, some operated by their trolley companies, some horse-drawn to clear sidewalks, and some motor-driven for street plowing; 54 had horse-drawn dump carts, but 51 also had motor trucks,

---

[44] *American City*, Sep. 1920, pp. 256–7; Feb. 1921, pp. 168–70; Oct. 1923, pp. 333, 353, 377–78; Aug. 1927, pp. 149–50; Feb. 1962, pp. 77–79; *New York Times*, Jan. 28 and 29, 1922; *Weather Record Book*, p. 74. Conrad P. Mook, "The 'Knickerbocker' Snowstorm of Jan. 1922 at Washington, D.C." *Weatherwise*, Dec. 1956, pp. 188–91.

and two used steam shovels to help in snow removal. Only two among those responding said they had snow-loaders, but already at least a dozen cities had acquired one or more Barber-Green loaders, among them Detroit, which marshalled an impressive arsenal of truck and tractor plows, belt and crane loaders, and dump trucks in its $121,000 bout with a record-breaking storm in 1925.[45]

A few rural towns such as Laconia, New Hampshire, and Geneva, New York, still used snow rollers to pack the snow for sleighs and sleds, but motorists generally bypassed such towns. As the flood of cars mounted, several northeastern states assumed the task of plowing and sanding state highways. Connecticut had taken the lead in 1917, followed by Pennsylvania, New Jersey, and Massachusetts in the East, and Colorado and Michigan in the West, while in New York the counties had shouldered that responsibility. The number of highway miles plowed totaled 27,000 in 1922 and would quadruple in four years. When a ten-inch snow blanketed the Pittsburgh area in January 1925, state plows helped to release cars stalled on the Lincoln Highway.[46]

But the state plows generally stopped at the city line. In that 1925 storm, the trolley company in Pittsburgh brought its thrity-one sweepers, nineteen scrapers, and two new snow-shovel cars into use in its efforts to maintain service. Its officials protested vigorously when a number of delivery wagons and trucks, attracted by the open car lanes, got their wheels stuck in the switches and obstructed the trolleys. By nightfall the street department was ready to tackle the problem of removing the snow from traffic bottlenecks. It marshalled a force of 450 men with thirty teams, sixteen dump trucks, two shovel plows, and a Barber-

---

[45] *Public Works*, 50 (Jan. 8, 1921): p. 37; Mar. 1923, p. 96; Dec. 1923, p. 386; Dec. 1926, pp. 410–12.

[46] *American City*, Oct. 1918, p. 289; Dec. 1920, pp. 583–84; *Public Works*, Aug. 6, 1921, p. 105; Oct. 22, 1921, p. 317; Jan. 14, 1922, p. 32.

Green snow loader for an all-night assault on the snow mounds. The "giant snow remover," as a *Post* reporter characterized the loader, had been acquired five years before, but had stood idle in its barn. It now demonstrated its ability to keep a string of dump trucks busy in removing the snow from busy intersections and crucial traffic arteries.[47]

As the cities that supplied the chief base of the increasing flow of traffic pressed the search for new methods of snow removal, old proposals for the development of snow-melting machines and the use of salt gained a new hearing. The South Park Commission in Chicago began experiments with a rotary-blade motor-plow modeled on those used by western railroads. The *American City* and other journals published extended descriptions of the various plows and loaders that gave promise of an effective urban response to the snow problem.[48]

But any who thought the challenge had been met should have visited Syracuse early in February 1925 after a thirty-six-inch snow buried the city under a frigid white blanket. "Overwhelmed by the heaviest snowfall in a full half-century," as the *Herald* reported, Syracuse saw its traffic gradually brought to a complete halt by 10 P.M. on January 29. Thousands of theater-goers and those working late were left stranded and had to seek shelter in hotels and rooming houses where every nook was soon crowded. Members of one theater crowd, who got out a few moments before the roof caved in from the weight of the snow, rejoiced in their good fortune, but many suffered real hardships plunging through waist-high drifts to reach home that evening. The girls at the telephone switchboard bunked down for the night in order to be on hand to open communications the next day. As Friday dawned with snow still blowing about in an otherwise paralyzed city, the

---

[47] *Pittsburgh Post*, Jan. 30, 1925.
[48] *American City*, Oct. 1923, pp. 347–53, 377–78, 604–06.

serious dimensions of the crisis began to appear. With hundreds of cars and trucks left stranded in the streets, it quickly became evident that early deliveries of milk and other essential supplies were impossible. Even more alarming was the discovery that the fire-fighting equipment, like all other equipment, was immobilized.[49]

Rail service, trolley service, mail service—everything was "paralyzed." That was the favorite word, but it was not quite accurate, for many were beginning to stir. At the behest of Mayor John H. Walrath, the public works commissioner hired seven hundred shovelers and sixty teamsters with heavy sleds to assist the city's twenty dump trucks in removing snow from James and Salina Streets and in opening a few vital intersections. Two of the city's five-ton tractor plows had broken down during the night, but a city crew soon had both in working order. With great effort they channeled a ditch down the center of several of the principal streets by the end of the second day. The *Syracuse Herald*, which brought out its afternoon edition that evening, had joined forces with the large Netherland Dairy and opened a free milk station at the *Herald* office to which parents could send for emergency supplies, as well as a copy of the latest news release. The *Herald* and its rivals as well as the recently launched radio station passed on the information that all schools including the university were closed "until travel became feasible again," and that more factories and stores had shut down for the rest of the week.[50]

For the average citizen or student at the university (where the author was then enrolled) it was an exciting holiday and long enough to really savor. By the end of the second day, most residents had opened paths from their front or side entrances out to the trench which many had cooperated in digging down the center of the street for

---

[49] *Syracuse Herald*, Jan. 30, 1925.
[50] *Syracuse Herald*, Jan. 30 and 31, 1925.

essential travel to other parts of town. A stroll along these trenches between five- and six-foot high snow banks, greeting strangers as fellow survivors and sharing the search for fuel or food or news, brought a new sense of community to many residents. News of the successful battle of the New York Central to push the Lake Shore Limited into and out of Syracuse only a half day late with the aid of eight extra engines, four pushing a plow in front and four dragging and pushing the train, provided a dramatic measure of the storm. The large inter-city trolleys finally opened their tracks with the aid of two steam plows borrowed from the Central, and the city trolleys got two lines moving by late Saturday. Most church services were suspended that day, but on Monday morning the church bells and factory whistles sounded an appeal by the Mayor and the Chamber of Commerce that all citizens get out and help shovel lanes sufficient for the resumption of traffic in major streets in every neighborhood. The urgent tasks of making deliveries of coal and food and collecting garbage could no longer be deferred.[51]

As the city struggled to resume its normal functions, it could not just return to its old practices, the *Herald* warned. A new and more effective snow-fighting program was needed, the Mayor agreed, and new provisions for the off-street storage of cars. One automobile dealer took the opportunity to point out that only enclosed cars such as he was promoting would serve as year-round cars in the future, and he invited prospective buyers to attend the auto show at the State Armory to view the new enclosed cars "all next week."[52]

---

[51] *Syracuse Herald*, Jan. 30 and 31, 1925; *New York Times*, Jan. 31, Feb. 4, 1925.

[52] *Syracuse Herald*, Feb. 2 and 3, 1925.

A SNOWBLOWER CLEARS THE ROCHESTER AIRPORT
(Courtesy of Division of Public Information,
City of Rochester, N.Y.)
As the motorized cities expanded and multiplied, linked by super highways and airlines, snowblowers replaced plows and dump trucks for many tasks.

CHAPTER FIVE

SNOW-FIGHTING IN MOTORIZED CITIES

1925–1955

The increasing flood of automobiles in the late twenties was transforming the life-style as well as the economy of American cities. Even towns of modest size found periodic wintertime interludes intolerable; more and more northern communities acquired mechanized snow-fighting equipment and adopted plans for its systematic use. Other technological innovations, most notably the development of air-transport services and the increased use of radio in communications, had significant influences on snow-fighting practices. The Weather Bureau established branch stations at the early landing fields and perfected the use of radio in advising aviators of weather conditions along their routes. In the early thirties, as radio acquired a popular following, the breakfast-time weather reports and occasional announcements of school closings further heightened public interest in blustery snowstorms. The local snow-fighting strategy became a subject of heated debate as motorists, discovering that slippery roads posed a greater hazard than snowdrifts, demanded a generous application of salt, cinders or sand, thus provoking loud protest from other citizen groups. As the automotive boom exploded, trolley companies converted to buses, and city after city faced the necessity of taking on the entire snow-removal problem, which was now seriously aggravated by the added obstruction of stranded or parked cars.

New inventions and new snow-fighting strategies restored the confidence in many cities, but mother nature retained her command. The snow problems were changing as all cities became motorized, and while both the vigor and the spread of the response were determined by the weight and scope of successive wintertime blasts, even cities that had been able to shrug off infrequent storms as historic anomalies were now forced to devise snow-fighting programs.

Thus it was not only the experience of heavy storms, as at Rochester in the early 1900s and at Syracuse in 1925, that produced action. Sometimes the report of a snow blockade in a neighboring city provided a sufficient alert, as Worcester demonstrated in the summer of 1925. An inland city of about the same size as Syracuse, with an annual snowfall that placed it with Portland, Maine, in close competition with the leading snowbelt cities, Worcester had relied for years on its heavy trolleys to open its principal streets. Clearly a snow deluge such as Syracuse had suffered would have paralyzed Worcester, too, and Leonard W. Russell, superintendent of its street department, formulated a new plan of action. Reviewing the city's experience, he determined that Worcester had to plan for five types of storms. A freezing drizzle could coat the sidewalks and hills with ice and required an immediate application of sand to reduce accidents. The trolley company could handle a light snowstorm with its rotary sweepers, and a heavier fall with its plows. The city had to be ready to bring out its scrapers and dump carts to clear the rest of the surface of the major streets in a third-rate storm, and it would have to re-double that effort when blizzard conditions developed. Finally, a heavy wet snow or an ice storm would require prompt city action to remove fallen wires and limbs. His first duty, Russell declared, was to consult the Weather Bureau in order to secure the best advice available as to the potential of an approaching storm and then to prepare to tackle it accordingly.[1]

Both Portland, Maine, and Albany, New York, with snowfall averages that placed them near the front line, had developed snow-fighting programs in the early twenties. Portland had spent $57,000 in a determined effort to open its streets during the harsh winter of 1922–23 when a total of 125 inches of snow (with 55 inches on the ground on January 17) challenged its survival. Most of that sum went to shovelers and cart men, but five years later the city tackled a series of storms that totaled 77 inches with a partially mechanized force and an outlay of $40,000. By 1931 it had two snow-loaders, two five-ton tractor plows, a dozen trucks with plows attached, and a score of motor dump trucks, which proved sufficient to handle a total fall of five feet of snow that winter with an expenditure of $68,000.[2] Albany, while continuing to require its trolley company to plow its tracks, acquired a second snow-loader and accumulated in four years an arsenal of snow-fighting machines that included six heavy-duty tractor-plows, fifteen trucks with plows attached, and a battery of motorized dump trucks, sufficient, it was hoped, to handle its worst storms.[3]

Despite the relief afforded to New York by its subways, its elevateds and its underground conduits, the great metropolis faced repeated snow emergencies in the late twenties. Six- and ten-inch snows posed serious problems there, forcing a closure of all schools on February 5 and 6, 1926, as the snow-removal forces, six thousand strong, battled to clear the streets. A year later a force of eight thousand tackled a Sunday blizzard and had most of the streets clear

---

[1] *American City*, Oct. 1925, pp. 389–92; Nov. 1925, pp. 503–06.
[2] *American City*, Jan. 1936, pp. 57–58; Feb. 1936, pp. 73–77.
[3] *American City*, August 1927, pp. 149–50; *Public Works*, Aug. 1928, pp. 296–99; *Portland Press-Herald*, Jan. 17, 1923.

for school and business on Monday.[4] The 1927 storm had hit Boston more severely and gave its school children a second snow holiday; it also brought appalling news of maritime tragedies, the worst since the great blizzard of 1888, as the *Times* reported. Philadelphia, while escaping the high winds that added to Boston's and New York's woes, received a deposit of ten inches of snow that clogged its streets for several days and forced the city to adopt a new snow-fighting strategy.[5]

Philadelphia, a city that had averaged only 24.3 inches of snow over the preceding forty years, scarcely half that of Boston and well below New York's thirty-inch average, had discovered that it could not rely on the sun every year. Having learned New York's lesson the hard way, it now determined to prepare to plow and clear a major portion not only of its business district but of its congested inner city as well. It acquired snow-loaders and dump trucks, tractor-plows and mechanical sweepers. And the next February, when an east-coast blizzard battered the New Jersey cities and clogged New Brunswick's Hadley airfield, Philadelphia, which largely escaped that storm, was able to send a small detachment of plows to help reopen the field.[6]

Pittsburgh, more accustomed to harsh winters, suffered an unusually heavy blow in February 1927 when a blustery storm dropped nineteen inches in its streets. By a vigorous use of plows and sweepers the trolley companies managed to maintain fair service, though several urban lines suspended operations. The city mustered a force of 640 to man its arsenal of plows, loaders, and dump trucks and sought to enroll a thousand more to help in clearing the

[4] *New York Times*, Feb. 5, 1926, pp. 1–8; Feb. 8, 1926, pp. 1–6; Feb. 11, 1926, pp. 1–8; Jan. 16, 1927; Feb. 20, 1927, pp. 1–6.

[5] *New York Times*, Feb. 5, 1926, pp. 1–8; Feb. 11, 1926, pp. 2–3; Feb. 27, 1927.

[6] *American City*, 28 Jan. 1928, pp. 103–06; Feb. 1928, pp. 111–14. See Philadelphia's map of snow plow streets; *New York Times*, Jan. 29, Feb. 19, 1928.

sewer drains to assure a quick run-off. In addition to the disruption of traffic, the city faced a hazard from the heavy weight of the snow on residential and business roofs, and ordered its fire companies to assist in removing the snow from several threatened structures. Hopes for an early thaw were moderated by fears of a flood, and meanwhile the council had to rush a supplemental appropriation to replenish the exhausted funds of its street-cleaning department.[7]

Several midwestern cities also received their share of snow blasts in these years. Buffalo, Milwaukee, and Chicago were particularly hard hit. Buffalo received more snow in each of three months in the late twenties than Philadelphia got in its worst year; one storm early in January 1928, which dropped thirteen inches of snow (moderate by Buffalo standards), hit with such force that the gale off the lake drove heavy wires into the city where plunging temperatures converted the slushing water into a thick coat of ice. Earlier snow-fighting strategies proved unsuited for such an emergency.[8] Both Chicago and Milwaukee suffered heavy deposits of snow in 1927, '28 and '29 from storms that ultimately hit several eastern cities as well. Chicago doubled and re-doubled its efforts and its outlays, which were more than $420,000 in 1919, the year of its heaviest snows, when its forces removed 387,412 cubic yards of it. Milwaukee, though less severely battered, increased its snow-fighting arsenal to include three snow-loaders, thirty-five motorized plows and numerous dump trucks—sufficient it was hoped to handle storms that rarely exceeded fifteen inches in depth.[9]

Several new snow fighters appeared on the scene in the

---

[7] *Pittsburgh Post-Gazette*, Feb. 21, 1927.

[8] *New York Times*, Jan. 16, 1927; Jan. 4, 1928, pp. 1–3; Local Climatological Data, Buffalo, 1900–1935.

[9] *Public Works*, Aug. 1930, pp. 19–20; *American City*, Aug. 1937, pp. 48–49.

late twenties. The city engineer of Duluth proposed the use of salt mixed with gravel to coat the crosswalks and hilly street surfaces in his northern outpost. Kenosha, Wisconsin, bought a snow-loader to assist its horse-drawn plows and carts clear the streets. Across the border, Toronto, which averaged less than half the snowfall of its neighbors south of Lake Ontario, decided in 1927 to launch its limited forces at the start of each storm rather than waiting until a heavy deposit could present serious problems.[10]

A new device known as the "snow fence", first developed along rural roads, was finding increasing use in metropolitan areas. The University of Wisconsin made a study of the proper design and use of natural and artificial fences. The new municipal airports, springing up on the outskirts of many cities, were among the first to use these fences as easy methods for trapping blowing snow at safe distances from the runways. Airports at northern cities also had to face the task of plowing and sweeping their newly-paved runways in the early thirties.[11]

The motorized urban society was becoming more rather than less sensitive to wintertime storms. A sleet storm in freezing weather, or even one or two inches of snow, presented real hazards to motorists, as the mounting accident rates in cities demonstrated. Moderate storms of four to six inches, once taken in stride by most cities, became major hazards, veritable "snow encounters", and brought new towns into the snow-fighting arena. Troy, which had long relied solely on its trolley company, acquired a snow-loader and plow attachments for several trucks, as it devised a plan to tackle its storms in pre-dawn engagements to clear the streets before the flood of workers appeared. Indianapolis, more concerned with slippery streets, acquired a lime

---

[10] *American City*, Dec. 1928, p. 72; *Public Works*, Aug. 1928, pp. 294, 324–25.

[11] *Public Works*, Aug. 1929, pp. 289–93.

The *Britannica* clears the ice-bound Boston Harbor, February 18, 1844
Courtesy of David Ludlum, Early American Winters, *Frontispiece*

The most frightening hazard from winter storms in Colonial and early American days was the havoc wrought among the sailing craft at coastal cities and later at lake ports. Thus it was at Milwaukee, battered by successive storms in the late 1860s, that Professor Lapham and Congressman Paine moved to persuade Congress in 1869 to establish a weather service to warn lake and sea ports of impending storms. By 1872 the Signal Service was issuing weather synopses to 55 weather stations, with maps and flags announcing prospective weather conditions.

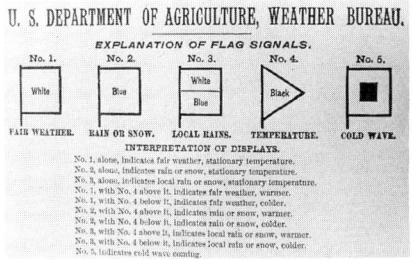

Weather Flags
U.S. Department of Agriculture

Quakers Sleighing near Philadelphia, 1800
Courtesy N.Y. State Historical Association, Cooperstown

A few sleighs appeared in Boston, New York and Philadelphia in the late 1600s, and Philadelphia required drivers to attach bells to their harness to warn pedestrians of their approach. Sleighs provided the major conveyance for wintertime travel and enjoyment until the 1840s and retained their popularity into the 1900s, longer than any other sport except skating. (See the Frontispiece.) Skating became increasingly popular as an individual pastime in the 1850s when sportsmen in snow-battered cities formed snowshoe and bobsled clubs and participated in inter-city contests. Several cities and towns staged winter carnivals; Albany's "Fort Orange" in 1888 was perhaps the most spectacular while the annual carnivals at St. Paul and Quebec were more enduring, and the snow forts erected in sideyards by hardy youths (including the author, eons ago) were innumerable.

Racing Sleighs on Mt. Hope Avenue in Rochester, 1950
Courtesy Rochester Museum and Science Center

Loaded Bobsled, Albany 1886
Courtesy N.Y. State Historical Association, Cooperstown

Snowshoe Club, 1860s
Harpers Magazine

Skating on the Genesee above the dam, 1860s
Courtesy Rocheser Historical Society

The Baltimore and Ohio railroad introduced the first plow, a horse-drawn contraption, to clear its tracks, in 1831. Soon other northern railroads were installing large V plows attached to the cowcatchers of their engines. Later, plow cars equipped with giant scoops topped by wedges, and propelled by two or more engines in tandem, were used. In the cities, horse-drawn plows pushed deep furrows into the adjoining path meant for sleighs and pedestrians. Companies, often the city's garbage collection unit, were assigned the task of shoveling and removing the snow to suitable dumps.

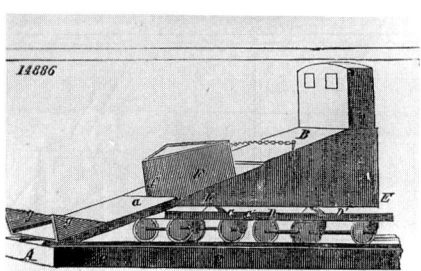

Baltimore's pioneer railroad plow
Sketched in *The American Journal of Science and Arts*, XX, July 1831

Three Engines Pushing a
Giant V Plow, 1850s
Harpers Magazine

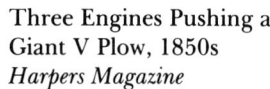

Improved Railroad Plow car (Patent Office)
Courtesy American City

Early horse-drawn Street Plow
Courtesy American City

Fire Engine in Snow Windrow in Rochester Blizzard, March 1900
Courtesy Eastman House

Denver's horse-drawn refuse dump carts at the snow dump in Blizzard of 1913
Courtesy Denver Public Library, History Department

V Plow on suburban line in Ohio
Courtesy Cleveland Plain Dealer

With the advent of electric cars, the trolley companies had to acquire plows that could be pushed around sharp corners. The original V plows proved unsuitable in city streets and were replaced by rotary plows and radial plows that cleared the tracks, but built high windrows that obstructed city traffic. New York and other cities experimented with melters and other devices in order to remove the snow.

Wilder Radial Plow Ruggles Rotary Plow, 1893
Courtesy Scientific American *Courtesy H. Dwight Bliss*

Denver trolleys caught in a deep snow, 1913
Courtesy Denver Public Library, Western Historical Department

Asphalt burner adapted for attack on snow and ice in an unsuccessful experiment in New York City, Feb. 4, 5, 1920
Courtesy Scientific American, Feb. 28, 1920

The Blizzard of 1888
Courtesy L. M. Strong, *The Great Blizzard of 1888*

In 1913 deep snow brought traffic to a standstill on 16th Street in Denver
Courtesy Denver Public Library

Under the recurrent blasts of harsh winters, each advance in the American industrial society called for new snow-fighting strategies and equipment: the burial of the festoons of electric wires in underground conduits; the replacement of horse-drawn plows, dump carts and shovelers by motorized plows, dump trucks and loaders; the fluctuating use of salt and other abrasives in the battle to maintain open and dry streets, producing in turn an environmental counterattack and new traffic and other regulatory controls.

A raging snow storm in November 1913 toppled several hundred telephone poles in Cleveland, blocking its streets
Courtesy Cleveland Plain Dealer

St. Louis first introduces a municipal dump truck into the snow-fighting fray
Municipal Journal, June 1914

Chicago clears its streets with a new snow-loader
Scientific American, Feb. 1921

A tractor rescues a stranded truck in New York, 1920
American City, Feb. 1922

A Barber-Green Loader at work in Milwaukee
American City, Oct. 1923

A Trip Loader, Tractor and Dump truck in New York City
American City, 1923

Syracuse tackled heavy drifts with its caterpillar tractor
American City, Oct. 1946

Rochester dumps snow through manholes into the Genesee
Courtesy Municipal Photographer

University Avenue in Rochester, Dec. 1945
Courtesy Democrat and Chronicle

Blizzards in a motorized age can be wonderfully beautiful, but desperately troublesome, trapping cars at the curbs, blocking highways and stranding travelers.

Blockade in Upper Manhattan, Dec. 12, 1960
New York Times, AP Wire Photo

The popular resort to salt to assure dry pavements has added a fleet of salt spreaders to the snow-fighting arsenals of many cities, and prompted the stockpiling of salt and other detergents. Counterattacks by environmentalists has led to a more cautious storage and distribution of the detergents. Snowbelt and other hard-pressed cities have learned to house their fleets in shelters equipped for quick dispatch.

Salt pile in Rochester, 1955
Courtesy, City photographer

Rochester's snow plows stored in a shelter equipped with an overhead trolley ready to lift and attach plows or blowers to the trucks as needed, 1955
Courtesy City photographer

Rochester's Fleet of Snow Blowers, ready for action, 1955
Courtesy City Photographer

Winter sports are rapidly growing in popularity. While ice sailing is still quite a rarity, snowmobilers are a common sight, and ice hockey and figure skaters vie for time on indoor rinks. Ski resorts and trails have multiplied, attracting thousands to favorite slopes and forests.

First Rochester snowmobile powered by a motorcycle engine, 1916
Courtesy Rochester Museum and Science Center

Ice Sailing Club, Minneapolis 1920
Courtesy Minnesota Historical Society

Memories of joyous sleighing parties on rural roads in past decades are stirred by this revival in a St. Paul park
Courtesy Minnesota Department of Natural Resources

Skiers make their way through snow-laden trees in Sugarbush Valley, Vermont
Courtesy Democrat and Chronicle, Nov. 24, 1977

Trail guides lead snowshoers through parks and forest reserves throughout the North
Courtesy Science Museum of Minnesota, 1979

spreader to distribute a mixture of sand and salt on icy roads. Rochester, startled by a series of traffic accidents during the first snowstorm of the year in November 1928, quickly scattered ten tons of salt and 688 tons of sand in January and February and ordered an increased supply for the next winter to safeguard against accidents. Akron, persuaded that salt would answer its rare and generally moderate snow problems more effectively than plows, became an outspoken advocate of chemical combat, but when widespread blizzards in the Midwest dumped large amounts of snow on the city in 1930 and again in 1935, it rushed to buy plows and dump trucks to clear its streets.[12]

Automobiles stalled in the snow-blocked streets or parked along the curb presented a new problem that plagued all northern cities in these years. A nine-inch snowfall on February 22, 1934, trapped thousands of cars in the New York metropolitan area, halting food deliveries, and prompted a new effort to carry milk and other necessary supplies by small planes. Merchants who stocked skis and snowshoes enjoyed a brisk market throughout the suburbs that year. Another heavy storm the next January re-emphasized the hazard of marooned cars and induced an increasing number of cities to ban overnight parking on streets served by snow-plows. A survey two years later by the American Public Works Association, a strong advocate of such a policy, revealed that although 73 of 94 cities banned all-night parking, enforcement was lax, and the problem was becoming more acute.[13]

Several major cities in the midwestern snow country devised plans to cope with storms of varied intensity and character. The frigid temperatures in Minneapolis saved it

---

[12] *Public Works*, Jan. 1932, pp. 42–43; *American City*, Oct. 1935, p. 13; *New York Times*, Mar. 27, 1930; Dec. 26, 1935, pp. 1–5; *Rochester Times-Union*, Nov. 26, 1928, Mar. 4, and Dec. 27, 1929.

[13] *New York Times*, Feb. 22 and 27, 1934; Jan. 23 and 24, 1935; *American City*, Oct. 1938, p. 11.

from many heavy mid-winter storms but made it especially vulnerable to November and March snows that required a battery of plows and scrapers to clear the streets before the deep freeze set in; meanwhile its park officials developed a flexible schedule of winter sports that took full advantage of the sub-freezing weather and made Minneapolis a "Winter Wonderland" for many adults as well as youngsters.[14] More prosaic Milwaukee, alarmed by a nineteen-inch fall in the March blizzard of 1930, which recalled its record snow of 20.3 inches in February 4–5, 1924, developed a proportionately larger arsenal of plows, loaders, and dump trucks to handle such emergencies. St. Paul was similarly prepared, and was in fact one of the first to acquire two of the new "Sno-go" machines that used rotating blades to rip into packed snow and throw it off to the side. With a full arsenal of motorized equipment, it spent an average of $60,000 annually during the mid-thirties on snow-fighting.[15] Denver, while reasonably equipped, was unprepared for a later September 1936 storm, which dropped 21.3 inches in twenty-four hours; the heavy wet snow caught the trees with leaves still on and brought down branches and wires in a tangled mass that blockaded most streets. Snapped telephone and electric wires silenced thousands of telephones and three of the four radio stations, and extinguished the lights and gas furnaces in many homes; even the fire-alarm system was disrupted for a time; but repair crews, numbering five hundred men and working day and night throughout the weekend, managed to restore service by noon on Monday. The trolley and street-cleaning forces kept a few lines in operation, but only the prompt cooperation of WPA relief administrators, who supplied some

---

[14] K. B. Raymond, "Minneapolis a Winter Wonderland," *Recreation* (Feb. 1946): pp. 590–92.

[15] *Public Works*, Jan. 1932, pp. 42–93; *New York Times*, Mar. 27, 1930, pp. 1–6; *American City*, Oct. 1937, pp. 61–63; Ludlum, *Weather Record Book*, p. 77.

three hundred trucks and over six thousand men, enabled the city to clear its streets by removing 23,000 loads of limbs at a cost of $210,000 during the seven-day siege.[16]

State and Federal Aid Enlisted

Denver was not the first city to secure federal aid in the form of WPA workers assigned to help break a snow blockade. Buffalo, after a harsh winter with storms in December and January that exhausted its snow-fighting funds and left its plows in disrepair, had suffered a third and record-breaking snow that swelled the fall for March 1936 to 38.5 inches and the total for the season to 103. Paralyzed by the weight of the storm, the city got only 250 men with ten plows into operation on the first day. The newly-elected mayor blamed the slow start on his predecessor who had let the motorized equipment deteriorate, but his pleas for $50,000 to re-activate the force produced only $25,000 from a divided council. Fortunately Erie County, supported by a state subsidy, got its crews out and began to open major highways into the city on the second day, when the Buffalo trolley company began to clear portions of a few lines with its heavy trolley-plows. The city sent 1,000 shovelers with two steam shovels and numerous dump trucks to tackle the drifts in lower Main Street on the third day. The widespread storm had aggravated a flood in the Ohio Valley, and as Buffalonians saw the federal government respond with relief funds there, the mayor asked and received WPA assistance in his battle to open the streets. With 8,000 and finally 10,000 relief workers on the job, assisted by a fleet of seventy-five trucks borrowed from the county, Buffalo broke its blockade on the fourth day, though many cars remained buried in

[16] *The Denver Post*, Sep. 27, 1936, pp. 42–93; *New York Times*, Mar. 27, 1930, pp. 1–6; *American City*, Oct. 1937, pp. 61–63; *Weather Record Book*, p. 77.

mounds at the edge of some streets until an early thaw released them. The depth of snow, together with the havoc wrought and the losses sustained, gave the St. Patrick's Day storm of 1936 a permanent place in Buffalo's history.[17]

The new federal involvement, as initiated by the assignment of WPA workers to snow removal at Buffalo, was a characteristic New Deal response during the depths of the Great Depression. Though only a temporary measure, it provided assistance to several storm-battered cities during a period of transition in snow-fighting from major dependence on manpower to a utilization of mechanical and chemical alternatives.

New York was one of the cities that benefited. Despite its massive transit systems, New York was still a walking city, and hundreds of thousands of pedestrians who used the sidewalks and crosswalks daily were adamant about prompt snow removal. The use of salt had been checked, and even the use of sand was questioned because of the danger of clogging the sewer pipes. Instead the city clung to its traditional strategy of making a prompt and mass attack on major snowstorms by providing a continuous stream of dump trucks to cart snow quickly to the East River docks. When a sudden drop in temperature turned a wet snow and slush streets into ice in January 1936, New York mustered twenty thousand unemployed men with picks and shovels to clear the pavements and purchased a "Sno-go" machine to attack the snow banks with rotating blades.[18]

The winter of 1936–37, while fairly mild throughout the Northeast, wreaked its vengeance on the West Coast. Portland and Salem, Oregon, received 16- and 20-inch snow-

[17] *Buffalo Courier & Express*, Mar. 18–22, 1936; *Buffalo Evening News*, Mar. 13, 1950.
[18] *American City*, Sep. 1933, p. 54; Jan. 1935; Jan. 1936, and Nov. 1936, p. 13; *New York Times*, Jan. 23 and 27, 1935.

falls respectively early in February 1937, more than double their annual averages. But it was Fairbanks in Alaska that took the brunt of the storms that dropped an all-time record of 65.6 inches of snow on that city in January, increasing its winter total to an unprecedented 134 inches. Accustomed to long but not hard winters, Fairbankers made a leisurely response, marveling at the dimensions of the snow and bringing out snowshoes to clamber over its drifts. Fortunately the wind, as in New York's great blizzard, had blown the snow into great piles on one side of many streets, leaving the other side relatively bare, which allowed residents to move about on foot. Unprepared for such depths, the city secured a loan of an Alaska Road Commission bulldozer and rotary-plow to break through the drifts, and pressed all its garbage trucks into service carting the snow away. An unusual January rain reduced the snow mounds somewhat but added to their weight. A second storm renewed the drifts and alerted business leaders to the need to remove the snow from the commercial streets to avoid a possible flood when spring arrived. By hard work and cooperative effort, the city and its residents were able to clear the principal streets on the twelfth day.[19]

New York and other experienced snow-fighting cities were developing a professional approach to the problem. New York organized a training course and took thirty selected men for a three-day briefing session to Lake Placid in the fall of 1937. On returning to the city these leaders met with 210 supervisors in a demonstration of the mechanical equipment on the newly-acquired landing field in Brooklyn. The cost of $8,000 on this training program was more than justified by the improved efficiency that winter. The department added a battery of ten rotary sweepers to its arsenal of snow loaders, dump trucks, and plows of various capabilities, but it still continued to rely on the army

---

[19] *Fairbanks Daily News,* Jan. 9, 16, 18, 19, 20 and 21, 1937; Ludlum, *Weather Record Book,* pp. 74, 76.

of shovelers, which was readily expandable in these depression years. To make the operation easier, it equipped its field inspectors with small addressograph machines attached to their belts to stamp the work tickets of laborers in the streets.[20]

The American Public Works Association devoted several pages to snow removal in its publication on *Street Cleaning Practice*. It gave major attention to snow-plowing, with diagrams and photos of several of the different kinds of plows, sweepers, and snow-loaders. An appendix recording the replies of 82 cities to its questionnaire disclosed that all but nine reported some snow-plowing, though 42 relied on their trolley companies for this service. Only seventeen reported efforts to remove the snow, and three of these relied on the trolley companies to do this job, too. Nearly half the cities used salt to some extent, and many of the others used sand or cinders. Only one city, Rochester, had undertaken the task of plowing all its sidewalks, but at least a score had now accepted snow-fighting as a municipal responsibility.[21]

Rochester felt the impact of technological advance in two unexpected ways in the late thirties. A decision by its transit company to shift from electric trolleys to busses on several lines turned responsibility for plowing thirty-seven miles of streets over to the city. Although the company continued to plow sixty miles, that transfer increased the municipal costs from $330,000 to $360,000 in 1937. Light snows the next winter brought a saving of $70,000, but a decision by the contractor who had plowed most of the sidewalks for many years to sell his horses and retire encouraged the city to buy a dozen baby tractors to perform that job. A review of these programs by the Rochester Bureau of Municipal Research late in 1936 and again in

---

[20] *American City*, Aug, 1934, p. 44; Feb. 1937, p. 11; Oct. 1938, p. 11.
[21] American Public Works Association, *Street Cleaning Practice* (Chicago, 1938) pp. 172–216, 373–75.

1940 suggested a few possible economies, but generally supported the acquisition of two additional snow-loaders, the employment of a crew of sanders to treat downtown crosswalks, and the opening of two new dumping chutes on river bridges.[22]

Edwin A. Miller, then in charge of the Rochester program, chaired a round table on snow removal at a convention of public works engineers in 1939, and advised urban snow-fighters to keep a step ahead of each storm. By checking storm possibilities with the Weather Bureau, by attaching plow blades to selected motor units ahead of time, by developing communication channels and staging practice alerts each November, he proposed to keep the Rochester snow-fighting forces ready for any emergency. Miller also made one more popular suggestion: that superintendents of snow removal should give a high priority to clearing the streets and entrances to hospitals and especially to churches where funerals were frequently performed.[23]

Rochester achieved an enviable record despite a limited use of salt in the late thirties. It again responded effectively to a blustery storm in January 1943; many schools in the suburbs made radio announcements of successive snow holidays, but the city schools were all open on the second day. A light winter in 1943–44 brought welcome savings, and although the city again budgeted $450,000 for snow removal, the sense of crisis had relaxed. Unfortunately, a heavy storm on December 11 and 12, followed by another that increased the month's total to a record-breaking December high of 41.8 inches, brought all traffic to a halt and presented Rochester's model snow-fighting forces with a back-breaking task. The city's eighty-five plows had barely finished opening a two-way channel

---

[22] *Rochester Democrat and Chronicle*, Sep. 17, 1937; Apr. 6, Sep. 23 and 27, 1939; *Municipal Research*, 1936, 45; Jan. 1940, p. 198.

[23] Public Works Engineers, *Yearbook* (1939). pp. 323–25.

through the principal streets when the second storm hit, filling the streets with new snow and marooning many buses. (The author was one of thousands who trudged for miles through the drifts to reach home during the second storm and in his case spent several days recuperating.) Plows struggling to re-open the traffic lanes had to settle at many points with a one-way passage, squeezing past cars buried in the drifts. With Herculean efforts, aided by a number of German prisoners-of-war who volunteered for jobs as shovelers, Rochester was able, despite new snowstorms in January, to get traffic moving in a fairly normal fashion by the end of the month. But already the opposition party was chanting, "In November, remember December." The incumbent Republicans survived that election, but when a sudden storm on November 29, 1945, dropped eleven inches of wet snow that again threatened a paralysis, the atmosphere at city hall became grim. Miller found another job, and the city, determined to keep the streets open whatever the cost, saw the snow-removal outlays soar to $834,949 for the season.[24]

Few cities had responded so dramatically to the snow problem. Both Syracuse and Buffalo, despite heavier deposits in these years, contented themselves with the purchase of a few additional plows, trucks, and snow-loaders to open the principal business arteries. Plowing and snow removal costs in Syracuse rose from an average of $50,000 annually in the early forties to $126,764 in fighting its 128.7-inch record-breaking snows of 1944–45. Stricken by its nineteen-inch fall of wet snow on St. Patrick's Day in 1936, Buffalo had purchased an additional snow-loader and five mechanical spreaders to supplement its plows; a generous use of salt had eased its snow-removal problem in the early forties. The two worst winters in the mid-

---

[24] McKelvey, *Snowstorms and Snow fighting*, pp. 15–17; *Rochester Democrat and Chronicle*, Jan. 5, 1943; *Rochester Times-Union*, Apr. 14, Sep. 27, Oct. 11, Dec. 12, 1944; Jan. 11, Nov. 30, 1945.

forties dispelled that complacency. With persistent efforts, Buffalo battled the repeated storms of December, January, and February 1944-45 that raised the season's total to 120 inches, but when a new storm dumped 36.6 inches on the city in mid-December 1945, boosting the total for that month to 51.1 inches, business leaders as well as suffering motorists demanded more forthright action. As a result, Buffalo spent over a million dollars on new snow-fighting equipment during the next two years. It increased its fleet of trucks with snow-plow attachments to one hundred, acquired two tractor bulldozers, four power graders, five sno-go rotary plows, and six large V-plows with wing-spread attachments—sufficient, it was claimed, to handle a sixteen-inch fall in one day. As an additional precaution, an enterprising trucking firm inserted some of the newly-devised radiant heating coils in its driveway to insure free movement in all weather. An adequate test of these measures was deferred, however, as several light winters enabled the "Queen of the Lakes" to clear its streets with the aid of salt spreaders, now twelve in number.[25]

While urban officials in northern cities were struggling to develop their snow removal plans and arsenals, bobsledders and sleighing enthusiasts, deprived of snow-packed roads, were taking to skiing, which enjoyed a sudden popularity in the 1930s. The sport, long practiced in the snow-covered mountains of Europe and already popular in Canada, now found a welcome base on the hinterland slopes of several northern cities. Heavy snows in New England and New York brought obscure hamlets in the forest-covered hills and mountains that fringed the Canadian border into view as skiing resorts which, with similar developments in the Rockies, attracted an increasing flood

---

[25] *Syracuse Herald*, Jan. 31, 1940; *Syracuse Post-Standard*, Mar. 10, 1945, *American City*, Mar. 1941, p. 11; 1945, p. 13; Dec. 1948 p. 13; *Buffalo Evening News*, Mar. 13, 1950.

of urban sportsmen seeking the new wintertime thrills.[26] Cold, dry winters continued to draw skaters and hikers to frozen lakes and rivers, but not without hazards, as the author discovered in January 1934. On a drive with my brother along ice-covered Susquehanna River near Harrisburg, the sight of numerous young men frolicking on its glistening surface enticed us out the next day to share the novel experience. Of course, a running river is not a still lake, and a protracted freeze gives it a variegated surface which can include long stretches of clear and smooth ice fringed by ridges of flaky and crushed slabs piled up by an early thaw. Arming ourselves with sturdy poles, we strode out on the ice, vying with some of the lads by taking a running slide on the clear stretches and testing the fringes with our poles. As we approached the center of the half-mile wide river, I took a running skid on a clear stretch of ice and, plunging my pole into its fringe, found myself standing in ice-cold running water up to my armpits. Fortunately, my brother, more cautious in his actions, was several steps behind and on solid ice. He quickly stretched his pole to me, and by grasping it and placing mine on the thin ice I had crashed through, I was able to climb out and start the long run back to his car and home where, in a tub of cold water, I managed to thaw out, reflecting, as I recall, on the loss that would have occurred of the fruit of my recently-acquired Ph.D. if the Susquehanna had been a half foot deeper. I have failed to find a tabulation of winter drownings, but they must have been numerous in cold, dry winters, though infinitesimal compared with the mounting toll of fatalities on urban streets and highways during winter storms.

---

[26] E. A. Bauer, *Crosscountry Skiing and Snowshoeing* (New York: Doubleday, 1975), pp. 21–22.

Snow-Fighting in Motorized Cities

An Increased Use of Salt

Both the plight and the accomplishments of the leading snowbelt cities became more widely known as the burgeoning radio networks broadcast news bulletins in the forties that increasingly featured Weather Bureau reports. The few widely-scattered severe storms of the generally moderate thirties—Denver's ice storm in September 1936, Fairbanks' great blizzard in January of '37 and Portland's ordeal that shortly followed—had been treated quite properly as anomalies. By contrast, the great storms in upstate New York in the mid-forties seemed much more real and deserving of close attention. The merits of prompt and forthright action were clearly apparent, but so were the costs and limitations of snow-fighting forces. as well as the economies promised by a use of salt when the snowfalls were moderate. And when cities with heavy snow records, such as Buffalo, made an increased use of salt, many officials elsewhere took note. Cleveland and Detroit, less frequently plagued by winter storms, had also turned to salt to keep their streets safe for motorists. Cleveland, in fact, had won a traffic-safety award partly because of the dry street surface it was able to maintain by a free application of salt and cinders. Detroit, an early salt user, had expanded its application and saw the costs of the annual snow-removal and salting program mount from $50,000 to $195,453 during the thirties. Schenectady and Springfield, Massachusetts, relative newcomers among the snow-fighting cities, both adopted salt as the principal agent in clearing their roads.[27]

The traditional hostility to the use of salt was disappearing as the concern for safe driving mounted. Dr. C. D. Hooker, research director of the International Salt

---

[27] *American City*, Nov. 1940, pp. 56–57; Sep. 1941, pp. 77, 83; Jan. 1943, p. 79.

Company, recommended the use of salt without any mixture on sleet-covered pavements, and on all streets when snow reached one inch in depth, and again at the end of a storm to keep the snow "mealy" for plowing. Its effectiveness was greatest at 30° and declined with dropping temperatures. Detroit, the automotive capital, perfected the salt spreader and employed it as a major weapon in its annual snow-fighting campaigns, which by 1945 called for an appropriation of $100,000 to start with, to be supplemented as needed. With an average of thirty-five salt runs during a season, the costs sometimes doubled that figure. Hartford, an old experienced snow-fighter, switched to salt, too, in 1943 to check its climbing ratio of accidents and was able in 1944–45 to clear its streets of a sixty-six-inch total at a cost of $88,582.[28] Philadelphia's Bureau of Municipal Research, in a review of the problems of that city, was baffled by the difficulty of planning its snow-fighting program. The mechanical equipment needed to handle an occasional thirty-one inches, as in 1940–41, would stand idle many years with eight- and ten-inch totals. The Bureau concluded that a judicious use of salt would enable Philadelphia to meet the challenge of hard winters more effectively.[29]

Heavy equipment was still in demand, however, All the new airports, particularly in northern cities, had to equip themselves to clear their landing strips quickly in order to maintain winter service. Thus the airport at Providence, not an aggressive snow-fighter, spent an average of $3,500 on this task annually in the early forties. New borderland cities purchased snow-fighting equipment to join the fray. Harrisburg, not too far north of Pennsylvania's southern boundary, had traditionally relied on the sun to clear its streets, but a hard winter in 1935–36 proved the folly of

---

[28] *American City*, Feb. 1939, p. 58; Oct. 1945, p. 103; Dec. 1946, p. 8.

[29] *American City*, Sep. 1941, p. 77.

inaction and persuaded the city to buy a quantity of plows and snow-loaders to guard against future snow crises. That investment paid off in mid-January 1945 when the city put its plows to work breaking channels through a record-breaking fall of twenty-one inches. Fortunately the state was now equipped to open the approaching highways and to help in clearing the streets around the capitol buildings.[30] Even heavy salt users, such as Detroit, required additional snow-loaders and dump trucks to remove the mush before it created new hazards in flooded underpasses, while of course the purchase of salt spreaders was a new financial burden on many towns.[31]

The shifting intensity of winter storms posed serious questions in many cities as to the merits of large investments in snow-fighting equipment. A succession of light winters permitted street-cleaning departments in some communities to economize by neglecting the maintenance of little-used plows and other machines. Chicago demonstrated the contribution that changing administrations could make to this neglect when a snow emergency early in 1948 urged its new officials to make a survey of the equipment of eighty mid-western cities. Much to its surprise, Chicago, the place where the snow-loader was first introduced, discovered that it was the only large city without one, and immediately purchased three to escape its reliance on shovelers. In New York State, where snow-fighting equipment was in demand every year, but not always in the same localities, it was finally decided in 1946 to abandon the practice of subsidizing county systems and to equip its Department of Public Works with plows and other machines that could be moved quickly into action in any snow-bound district.[32] Iowa also had a contingent of

---

[30] *American City*, Feb. 1941, p. 41; Jan. 1944, p. 49; Ludlum, *Weather Record Book*, p. 76; *Popular Science*, Jan. 1947, p. 79.
[31] *American City*, Oct. 1945, p. 103.
[32] *American City*, Jul. 1946, p. 11; Feb. 1948, p. 11.

state highway plows, but Des Moines relied on its city plows and dump trucks to clear its loop district after a raging New Year's Day blizzard in 1942 dumped a record-breaking 24.5 inches on its streets. When even some of its plows became stalled in the drifts, an amused reporter noted that among the many other marooned vehicles was one bearing the official markings of the coast guard. With determination, the city crews restored essential traffic by the third day, winning applause for their efforts.[33]

But the trend was toward an increased use of salt, or salt mixed with cinders or sand. Even New York City, long opposed to its use, quickly bought a fleet of sixty salt spreaders; by the middle of the decade it was annually purchasing five hundred carloads of rock salt from the Retsof mine upstate and candidly described salt as its most valuable snow-fighting weapon. Newark, across the Hudson, announced its plan to salt all storms up to four inches and to hold the plows back until seven inches had fallen. Barre, Vermont, cut its snow-fighting costs in half by the use of salt. Baltimore, disillusioned by the accumulation of dirt resulting from the use of cinders on icy pavements, finally repealed an early ordinance prohibiting the use of salt and purchased four salt spreaders.[34]

The increased use of salt precipitated a debate over the damages it caused, particularly to cars. Carl D. Warner of Detroit labeled the charge "imaginary". The chief object was safety, he declared, citing the success Detroit had achieved in the reduction of accidents during the ten years when it had made a generous use of salt. He listed a dozen cities which had increased the use of salt in the mid-forties, boosting the tonnage used to over 1,000 in Cincinnati, to almost 2,000 at Cleveland, and to 35,000 in New York City.[35]

---

[33] *Des Moines Tribune*, Jan. 1, 2 and 3, 1942.
[34] *American City*, Aug. 1944, pp. 64–65; Feb. 1947, p. 69; Oct. 1947, p. 99; Jan. 1948, p. 83.
[35] *American City*, Aug. 1947, pp. 92–93.

Statistical data showing the increased use of salt failed, however, to convince some critics, and a new search commenced for a substitute that would be less corrosive. Cinders, gravel, and sand had proved objectionable on city streets because of the accumulated dirt and the damage to sewers. Salt mixed with such abrasives lessened that hazard but limited its effectiveness. City chemists at Rochester, Pittsburgh, and Kansas City proposed the use of varied additives to reduce the corrosive effect of salt, and in 1948 Akron joined in an experimental use of an additive called "Nalco" designed by John Temmerman, Rochester's city chemist. Within two years the number of cities purchasing one or another of several additives had mounted to one hundred, despite the increased cost, and the claims of their advocates were eagerly reported.[36]

Raging and Widely Scattered Blizzards

Salt of course was not an effective response to major blizzards. The snowstorms that battered several upstate and midwestern cities in the mid-forties, promoting a greater use of salt there, were followed in 1947–48 by harsher storms over eastern New York and New England that revived memories of the great blizzard of 1888. The great snowstorm that hit New York City on December 26 and 27, 1947, actually exceeded the blizzard of 1888 in the depths of its deposit—26.4 inches—and in the losses it inflicted. The city was of course better prepared to cope with it, and spent an estimated $6,000,000 in combating its fury; the damages inflicted on the metropolitan economy were nevertheless staggering. Although the major rail lines were now equipped with plows and blowers to maintain

---

[36] *Rochester Democrat and Chronicle*, Feb. 20, 1946; Dec. 27, 1947; *American City*, Dec. 1948, pp. 86–87; Jan. 1950, p. 21; Sep. 1950, p. 133.

service, one commuter train remained stalled for several hours on the Long Island meadows, and the number of commuters jamming the Grand Central and Pennsylvania stations, waiting for delayed departures, created near mob scenes as recalled by a correspondent who experienced the frantic crush. Subway and elevated trains continued to operate, and the city's snow-plows managed to open lanes in some major streets, but thousands of cars were stalled in the drifts and many streets were impassible. All air traffic was suspended for two days, and milk and food deliveries had to be made by sled or toboggan in many districts. After laboriously freeing five thousand cars on the first day, the police ordered all private cars off the streets. When the mayor joined with other leaders in appealing for the aid of all residents in clearing the streets, health officials warned shovelers to take it easy. Many residents and visitors alike enjoyed the unexpected holiday; some, after a thrilling stroll down the center of normally congested avenues now completely deserted, hurried indoors to share their experiences with distant friends by telephone, loading the exchanges with a record sixteen million calls that day.[37]

Great storms stirred new interests among amateur weather watchers as well as professional meteorologists. It was a coincidence that the new journal, *Weatherwise*, made its appearance in February, two months after the record-breaking snows of December 1947, but the first issue carried an excited report of that storm sent in by a youthful member of the recently organized Amateur Weathermen of America for whom the magazine was originally designed. The new journal featured articles on current experiments with rain- and snow-making, short-term and long-range forecasting, and the new efforts to probe the heavens with the aid of meteorological balloons, planes and

---

[37] *New York Times*, Dec. 27, 28 and 29, 1947; Eugene Kinkead, "New York City Weather Extremes," *New Yorker*, (Jan. 31, 1977): p. 58.

rockets. But snow was the absorbing hobby of its editor, David M. Ludlum, and his second issue contained a meteorological report on the "East Coast Storms" of December 26–27, 1947, by Professor James E. Miller of New York University. Miller's article set a pattern of scholarly analysis in a form comprehensible to laymen that would win the new publication a respected following among weather-watchers throughout the country.[38]

The snowstorms of the mid- and late-forties were also promoting new snow-fighting techniques. Hartford, which had suffered record-breaking snows, totaling 45.3 inches in December 1945, had developed an efficient snow-fighting force and, with the timely advice of a private forecaster, battled the December blizzard two years later with self-assurance. An acute shortage of chains for automobile tires there and at Boston, hit by the same blizzard, symbolized the new demands of snowstorms in the motor age. They also brought the use of salt more prominently into the snow-fighting strategy of Boston, Hartford, and even Cambridge and New York City.[39]

When the cold winds shifted to the west and dropped heavier snow deposits on Omaha, Milwaukee, Pittsburgh and Cleveland, doubling their annual averages for one or two seasons, advocates of a generous use of salt and demands for more powerful plows found a hearing there as well. Omaha received its heavy blow late in January 1949, the climax of widespread blizzards that dropped one to two feet of snow over most of Nebraska. For Omaha it was not as heavy as a storm in March 1923, but that snowfall had ended a drought in the area and "worth a million dollars." In contrast, this January blizzard trapped thousands of cattle still in the fields, isolated ranch families, and

---

[38] *Weatherwise*, Feb., Apr., 1948, pp. 5, 36–38.
[39] *American City*, Dec. 1946, pp. 80–81; Aug. 1947, pp. 92–93; Oct. 1948, pp. 88–89; Jan. 1950, p. 21; Sep. 1950, p. 133; Nov. 1950, pp. 95–98; Nov. 1956, p. 129.

threatened the economy of the entire region. An appeal by the governors of three states brought an order from President Truman sending army trucks, bulldozers, and helicopters to help open the roads and relieve the stricken area. Omaha received some assistance from National Guard units but vigorously marshaled its own forces— seventeen tractor-plows, forty dump trucks, and three new snow-loaders—in an around-the-clock attack on the drifts produced by twelve inches of snow. All schools closed for the duration, and the mayor appealed to drivers to equip their cars with chains and not to park on snow-plow streets. On the second day, five construction firms supplied crews and equipment on the second day to help the city crews battle the storm in twelve-hour. Most streets opened two days later. To guard against future crises, Omaha Steel Works began to speed its production of snow-fighting equipment.[40]

January 1950 delivered its harshest blows at West Coast cities. Frigid winds from Canada plunged temperatures at Portland, Spokane, and Seattle to record monthly lows and brought snowfalls that also established new monthly records. In Seattle, which received the heaviest snows, the transit system had its salt and sand spreaders out on January 2nd, determined to keep traffic moving. Although the cost was burdensome, as its managers admitted, the increased revenues from passengers who usually drove to work, more than offset the loss, and the officials announced alternate routes for a few lines on hills that proved impassable. When a still harsher blizzard hit ten days later, dumping twelve inches of new snow in gales that piled drifts up to six feet in some streets, the city pressed its nineteen snowplows and numerous street scrapers into an around-the-clock battle to keep the major arteries open. As the storm continued, prompting the local weather station to pronounce it the

[40] *Omaha World Herald*, Mar. 15, 16 and 18, 1923; Jan. 28, 29 and 31, 1949.

Snow-Fighting in Motorized Cities

worst in Seattle's history, the transit system had again to revise and curtail its routes, but managed to maintain essential service. The schools remained closed for a full week, but the airport re-opened one runway on the fifteenth, when trains and busses also began to move again. But with frequent squalls, the deep freeze continued, and it was not until February 1, that the Seattle *Post-Intelligencer* could confidently declare, in a banner head on its "Sunrise Edition," "Let's Forget Weather and Get Back to Normal."[41]

Several midwestern cities were more heavily blanketed in the early fifties. Minneapolis received record-breaking snows, totaling forty inches in March 1951, and Milwaukee 30.7 inches that December, but the storms were protracted and these cities were prepared to cope with them. All schools and many factories in Pittsburgh, hit by a record-breaking twenty-eight inches of snow late in November 1950 before the salt trucks were fully ready, had to close and the traffic blockade continued for three days. Governor James B. Duff declared a state of emergency, closing all banks and mustering the National Guard to maintain order in the Golden Triangle where all stores and offices had suspended operation. The mayor warned motorists to stay off the streets unless equipped with chains and urged non-essential workers to remain at home. Efforts to recruit volunteer work crews to open lanes in neighborhood streets produced some results, but the National Guard had to bring out its army trucks to break through the drifts in some districts to open a path for ambulance and fire trucks. The trolley company finally freed all of the 150 street cars trapped at the height of the storm, and the arrival of a detachment of tank trucks from Ohio broke the threat of a milk famine on the third day. The city council, which made an emergency appropriation of

---

[41] Ludlum, *Weather Record Book*, pp. 55, 76–77; *Seattle Post-Intelligence*, Jan. 4, 5, 14, 15, 16, 27, 28, 30, and Feb. 1, 1950.

$250,000 during the early stages of the storm, learned three weeks later that the costs had "snowballed" to $600,000. Cleveland, with a record twenty-one inches on the ground on November 27, experienced similar losses from the same storm. Again the governor declared an emergency and called out the National Guard to discourage looters. The Guard used Sherman tanks to help move some of the 20,000 stranded automobiles. If the snowfall was slightly less than that of the city's paralyzing blizzard in 1913, the loss of business was much greater, and many criticized the city's hesitant use of salt.[42]

Only a few cities resisted the increased use of salt. Minneapolis and St. Paul, partly because their frigid temperatures reduced its effectiveness, made little or no use of salt and concentrated on the development of efficient mechanical devices. Memphis, well south of the snow country, relied chiefly on the sun to remove its infrequent falls of snow. A heavier fall, such as the twelve inches of January 18, 1948, brought its trolley-plows and street sweepers into operation to clear a few key routes, but these rare storms served chiefly to stimulate old-timers to recall the record eighteen-inch storm of 1892, or the blizzard that had battered Memphis on December 7, 1917, stalling all traffic for two days. Milder weather soon came to its rescue and Memphis escaped the need for salt.[43] Portland, Oregon, alerted by an unusual succession of storms in January 1950, found it possible to open its streets with the aid of plows attached to garbage trucks and the use of shovelers to clear the sewer drains. Salt, the city's horticulturally-minded leaders maintained, would prove more damaging

---

[42] *New York Times*, Nov. 25, 26 and 27, 1950; Mar. 20, 1951; *American City*, Nov. 1953, pp. 107–08; *Cleveland Plain Dealer*, Nov. 27 and 28, 1950; *Pittsburgh Press*, Nov. 27, Dec. 20, 1950; Charles L. Bristor, "The Great Storm of November 1950," *Weatherwise* (Feb. 1951): pp. 10–16.

[43] *Memphis Press-Scimitar*, Mar. 17, 1932, Dec. 23, 1963; *Memphis Appeal*, Mar. 17, 1931; Dec. 7, 1935.

than beneficial in the "Rose Capital." Champions to Washington's parks were also becoming doubtful of the merits of salt in the national capital's snow-fighting strategy, but a second objection there sprang from the short circuits it caused on the underground third rail used to power its trolleys. Yet despite these criticisms, a new survey of the snow-fighting programs of American cities found an increased application of salt in the eighty-two cities that responded, but also an increasingly vocal opposition to its use.[44]

In the leading snow-fighting cities, all weapons were considered legitimate if not always necessary. New York, the best equipped of all and still experimenting with melting devices, had now accepted salt as an essential aid. Indeed for several years after the record storms of 1947–48 it had kept its plows and loaders in their barns, relying exclusively on salt and flushers to wash away the light snows. It was not until January 10, 1954, that a seven-inch fall, the heaviest in five years, brought its mechanical equipment and a force of fourteen thousand shovelers again into the fray.[45] Similarly, Buffalo, after upgrading its snow-fighting equipment following the hard winters of the mid-forties, placed its chief reliance for several years on salt. In November 1950, however, and in December 1952, and March 1954, it would bring its snow-fighting forces out in full force in the suburbs, and now also with state crews on the highways. But in the early months of 1948, Buffalo was content to despatch its salt spreaders on frequent runs and scattered 6,000 tons to reduce the hazard of fatal and otherwise destructive skids. Even Syracuse, used to heavy snows, found it possible to open a wider spread of streets by a greater use of salt and scattered

---

[44] *New York Times*, Mar. 20, 1951; *American City*, Oct. 1953, p. 127; Jan. 1955, pp. 104–05; Sep. 1955, p. 158.
[45] *New York Times*, Jan. 10, 12 and 15, 1954; T. Napier Adlam, *Snow Melting* (New York: Industrial Press, 1950).

9,000 tons of it during the 100-inch season of 1954–55.[46]
Rochester received prods from both directions. It was currently enjoying a succession of relatively light winters, with annual snowfalls ten inches below normal, but like Buffalo, it was discovering that light snows spread over a period of four, sometimes five, months required more rather than less salt to maintain safe driving. Machine mixing of 15,000 tons of rock salt with 250 tons of the inhibitor, now imported in bulk from Chicago, raised the total cost of the salting program in 1951–52 to $190,000. An unusually mild winter, with snowfalls limited to two months early in 1953, allowed the city to cut its salting runs from an average of seventy-nine to thirty-four and to save half of its stockpile of salt and reduce its outlays proportionately. The reduction, however, was only temporary, and later that year the council voted to purchase thirty-six new plows and a half dozen tow trucks to remove cars illegally parked in plowing lanes or left stranded in the snow drifts. These costs raised the total for salting and snow removal to $950,000, and a further $150,000 for sidewalk plowing to be charged to assessments. It was not the Blizzard of 1888, but the hard winter of 1944–45, that Rochester could not forget.[47]

---

[46] *American City*, Dec. 1948, p. 13; Dec. 1955, pp. 76–77; *New York Times*, Nov. 27, 1952, Mar. 30, 1954; Local Climatological Data, Annual Summaries, Buffalo and Syracuse; *Buffalo Courier and Express*, Nov. 30, 1950.

[47] McKelvey, *Snowstorms and Snow fighting*, pp. 16–19; *Rochester Democrat and Chronicle*, Dec. 9, 1951; *Rochester Times-Union*, Apr. 8, 1953; Jul. 27, 1954.

A SKYWAY IN MINNEAPOLIS DEFIES THE SNOW
(Courtesy of the World Affairs Center, University of Minnesota) As municipalities accepted the task of clearing the streets, merchants in the business districts of thriving northern cities constructed skyways to assure the comfort of downtown shoppers and workers.

CHAPTER SIX

SNOW-FIGHTING IN A METROPOLITAN ERA

1955–1966

In the late fifties and early sixties, more cities stockpiled salt and tooled up to combat blizzards and ward off snow blockades. Blustery snowstorms continued their irregular, almost unpredictable course, hitting cities in the snow country repeatedly and throwing occasional punches at their neighbors to the south and west. Lighter snows blanketed wide areas as in the past, but their inhabitants could no longer shrug off even a three- or five-inch snow as wintertime "landscaping". The cities and towns had grown into sprawling metropolitan districts, dependent on motorized transportation that often chugged to a halt at the onset of a storm, or slammed into chain-reaction pile-ups. Cities equipped to clear a business district of modest size could not cope with a snowfall that clogged streets throughout a metropolitan area. Experienced snow-fighters such as New York City, determined to remove the snow even from residential districts, faced mounting costs that strained their fiscal resources. Merchants and home owners had to buy plows, or contract with service-men to clear their driveways and parking lots. The compulsion to get the ever-increasing and now completely motorized traffic moving quickly after the onset of a storm presented unprecedented difficulties.

Snow in the Cities

Technological Innovations

The mounting challenge produced several technological innovations and shifts in strategy. Efforts to improve the mechanical equipment made some striking advances. The use of salt increased so dramatically that it aroused opposition from new groups of environmentalists who joined motorists protesting the corrosion of their cars. As the combined costs for mechanical equipment and salt reached staggering heights, cities and towns appealed for state and federal assistance, and explored new schemes for metropolitan cooperation. And as private motor cars competed increasingly with public transit facilities, the struggle to keep the streets open during snowstorms forced the cities to ban parking on snow-plow routes and to tow marooned cars. Efforts to anticipate these problems produced new traffic regulations, new definitions of snow emergencies, and exciting new improvements in weather forecasting. Although an occasional official, proud of his city's snow-fighting program, declared confidently "Let it snow" or "we have licked the problem," as in Wilkes-Barre, Pennsylvania, in 1957 and in Traverse City, Michigan, four years later, many of their fellow superintendents were ready after the heavy snows of the early sixties to gather at a series of annual Northeast Conferences on Urban Snow Removal in an earnest quest for new procedures.[1]

Except for the leading snowbelt cities, Buffalo and Syracuse, which, with Erie, Pennsylvania, received heavy snows in November and December, the winter of 1955–56 was relatively mild in major cities throughout the country, until the middle of March when a widespread blizzard blanketed the Northeast. From Rochester and Syracuse eastward through Albany, Worcester, and Concord, the

[1] *American City*, Oct. 1957, p. 143; Jan. 1962, p. 97; Jun. 1963, p. 162.

snow blanket deepened until it was more than 43 inches for the month at Hartford and 46 at Portland. Streets and highways throughout the vast area were dotted with abandoned cars obstructing the efforts of urban and state plows to re-open them. Rochester and Syracuse, with newly improved equipment, tackled the storm effectively, but Hartford and Portland were paralyzed and remained immobile for three days. "Old timers [in New England] never saw anything like it," *Weatherwise* reported. Radio broadcasts announced the closing of schools, libraries, airports, and factories; public health authorities attributed 164 deaths to the storm.[2]

Yet it was New York and Boston on the eastern fringe of the storm, with thirteen and sixteen inches of snow respectively, that suffered the greatest injury. As the daily flood of cars caught in these cities at the onset of the storm struggled to get home, thousands became marooned and had to be abandoned. Tagging the obstructing cars served little purpose; many had to be pushed aside or towed away to clear the streets. Libraries and museums as well as schools closed; funerals were postponed. In New York City the sanitation department had eight thousand men and two thousand pieces of equipment at work on the second day. Mayor Robert Wagner, while pressing the recruitment of additional workers, hoped to keep the expense under the $7 million cost of the great storms of 1947–48. With great effort the city's heavy plows opened lanes in most streets by the close of the second day. Buses as well as elevateds resumed operation, relieving congestion in the subways. Telephone lines, now widely buried, enjoyed unprecedented business, and WNYC announcers broadcast successive lists of closed institutions and factories

---

[2] *New York Times*, Mar. 19 and 20, 1956; *American City*, Dec. 1955, p. 76; *Rochester Times Union*, Mar. 20, 1956; Ludlum, *Weather Record Book*, pp. 74–75; David Ludlum, "Some Notable Snows of the Winter 1955–56," *Weatherwise*, Feb. 1957, pp. 26–28; *Portland Press-Herald*, Mar. 21, 1956.

and suspended events. By the third day, with losses estimated at $150 million, the great metropolis began to return to normal.[3]

Reports of the ravages of the storm boosted the sales of snow-fighting equipment. A correspondent of the *Times*, after an in-depth survey of the subject, estimated the total sales for the year at $17 million. The principal demand, as reported by eight competing firms in a half-dozen states, was for big plows for city, state, airport, and railway use, but a growing market for small plows and blowers, reflecting the new demand for equipment to help clear private driveways. Even the traditional snow shovel was a "hot" item as the sales for the year reached one million.[4]

The competing firms were constantly improving their offerings. Blowers with lift shovels, reversible, tripping plows displacing the earlier V plows and front-end loaders vying with the old conveyor-belt type. New speed-plowing was transforming the strategy of road clearing, and new salt spreaders were assuring a more even distribution. The demand for snow-removal equipment was spreading to include the new suburban shopping centers and the industrial parks, both of which had to clear vast parking lots. But the airports, metropolitan parkways, and new superhighways needed the most attention and brought a planned coordination of salting, sanding, and plowing programs that pushed salt again into the controversial limelight.[5]

Salt users were increasing in number and becoming more outspoken in their claims. Chicago ordered its salt by shipload, 5,500 tons in a load, and prepared to treat 800 miles as conditions required. Wilkes-Barre as well as Pittsburgh turned to salt to reduce accidents and hopefully

[3] *New York Times*, Mar. 20 and 21, 1956.
[4] *New York Times*, Jan. 18, 1957.
[5] *American City*, Nov. 1956, p. 129; Jan. 1957, p. 14; Oct. 1857, pp. 106–09.

costs. But Rochester, which had succeeded in bringing down the number of accidents, saw its salt bills and other costs continue to rise, while the protests against salt corrosion likewise mounted. The Rochester department's effort to economize by skimping on the use of Temmerman's expensive inhibitor was reversed as the city endeavoured to keep its streets clear.[6]

That decision was soon to be tested, as were the strategies of the other snowbelt cities, by the onset of a record-breaking snowstorm in February 1958. A storm on January 19 had dropped fourteen inches on Rochester, and a second storm on February 5 dropped more than a foot. Then, a third and harsher storm deposited another 13 inches on February 16 and continued for four days, until the total for the month reached an unprecedented 64.8 inches. Buffalo, sixty miles to the west, got 54.2 inches that month, while Syracuse, eighty miles to the east, received a total of 72.6 inches from the successive storms and shivered under a white blanket that averaged four feet on February 19. Syracuse declared a "snow emergency" under a new law that permitted municipalities to assume additional powers in crisis situations. As the storm spread into the Albany area, Governor Harriman announced a snow emergency for the upstate region and requested federal assistance similar to that accorded flood-devastated districts. Although Washington turned a deaf ear, the state sent highway plows to relieve some of the towns under siege, and used helicopters to rush needed supplies to snow-bound victims in isolated hamlets.[7]

The cities, for the most part, had to battle their drifts themselves. Rochester, which now housed its plows in

[6] *American City*, Feb. 1956, p. 103; Oct. 1957, p. 143; *Rochester Democrat Chronicle*, Jan. 5, 1957; *Rochester Times Union*, Feb. 1, 1957; Aug. 28, 1957.

[7] *Rochester Democrat & Chronicle*, Feb. 18, 1958, pp. 1–7; *New York Times*, Feb. 5, 1958, pp. 26–9; Feb. 11, 1958, pp. 32–33; Feb. 17, 1958, pp. 1–8; Feb. 21, 1958, pp. 1–6; *Syracuse Post Standard*, Feb. 17 and 18, 1958.

municipal garages ready for prompt action, managed to keep its streets open for emergency use throughout the protracted three storms, but bus traffic was slowed and all schools and libraries were closed. Syracuse suffered a longer suspension of services, and while its snow-fighting forces battled to open the streets, the editor of the *Post-Standard* found occasion to reminisce on the earlier storms of January 1925 and March 1956.[8]

Big storms were accepted as part of the Syracuse experience, but in New York City, where the snow measured only 8.5 inches, the response was more vigorous. The city had its salt spreaders out early and soon moved into action with plows, loaders, and dump trucks, helped this time in the suburbs by a contingent of National Guard trucks assigned by the governor to snow-removal duty. New York State set another precedent this year when it leased a number of snow-plows from Ohio, which was enjoying an open winter. Nearly $1.7 million was spent on snow-removal activities upstate and in the metropolitan district; more than twice that amount was spent by New York City.[9]

In Rochester, despite the excessive depth of snow, a generous use of salt at the start of the first storm and before the second and third storms hit, had eased the plowing and removal operations. Some motorists, fearing damage to their cars from the salty slush, again protested its excessive use. A National Association of Corrosion Engineers proposed a test of the costly inhibitor in use at Rochester. It sponsored a program to attach metal tags to several thousand cars in that city and to a similar number in Syracuse and Buffalo, which normally used less than half Rochester's salt tonnage and no inhibitor. After an

[8] *Rochester Democrat & Chronicle*, Feb. 18 and 19, 1958; *Buffalo Courier-Express*, Mar. 23, 1958, pp. 38–42; *Syracuse Post-Standard*, Feb. 10, 1958.
[9] *Rochester Democrat and Chronicle*, Aug. 6, 1958; Feb. 25, 1959; *Rochester Times Union*, Feb. 14 and Jul. 3, 1959.

exposure of six months during another hard winter, the tags were re-assembled in February 1959, and to the gratification of John Temmerman, demonstrated the great merits of his inhibitor. Rochester accordingly increased its orders for salt—and the protests against its use grew louder.[10]

The harsh winters of 1957–58 and 1958–59 brought several new bidders into the market for rock salt. Mayor Richard Lee in New Haven, after battling in vain to open the city streets blocked by twenty inches of snow in January and February 1958, decided to use a mixture of salt and sand the following winter and also to ban overnight parking down town. A second edition of the American Public Works Association manual on *Street Cleaning Practice*, published at Chicago in 1959, gave much more attention than its predecessor to the use of salt; it cautioned, however, that salt was of little use at temperatures below 10° and completely ineffective below 6°. Among the cities polled, several new salt users appeared, even including St. Paul and Portland, Oregon.[11]

But salt would not do the job for a community hit by a snow avalanche such as Oswego suffered in the winter of 1958–59. Its only hope was for some outside assistance. When the first heavy snow fell in December 1958, Rochester, some seventy miles to the west and confident of its equipment, rushed a contingent of plows to help Oswego break through its drifts. The dramatic gesture prompted the state to promise a battery of plows, too, and Rochester was able to withdraw its force in time to be ready for another snowy siege, which broke all previous January records in that city. But while Rochester kept traffic moving in a 140-inch season that took second place in the city's

---

[10] *Rochester Democrat and Chronicle*, Aug. 6, 1958; Feb. 25, 1959; Rochester *Times Union*, Feb. 14, Jul. 3, 1959.

[11] *American City*, Nov. 1958, pp. 128–30; David Ludlum, "The Hard Winter of 1958–59," *Weatherwise* (Feb. 1960): pp. 26–28.

snow records, Oswego continued to flounder in a succession of snows that totaled over fourteen feet.[12]

The harsh winter of 1958–69 proved to be only a foretaste of impending furies. Denver received another early blow on September 29, 1959, with fourteen inches of wet snow and a tangle of tree limbs, wires, and abandoned cars making streets impassible for three days. Because of its increased size, the city's losses soared above the $7 million suffered in the record-breaking September storm of 1936. Three months later a "devastating" ice storm hit the Rochester area, felling back-yard power lines and cutting off electricity and gas in thousands of homes throughout the city and its suburbs. Rochesterians were accustomed to severe storms, but few of the chilly revellers at darkened New Year's parties realized that they were dancing on the tip of an iceberg.[13]

Weather-watchers and the public generally had long been aware of the inscrutable character of many winter storms and of the hazards they posed. Many had backed the efforts of the Weather Bureau to perfect its observation techniques and to improve its storm-warning services. The announcement in January 1959 that a new satellite, *Vanguard II*, had been successfully launched and that it would supply climatological data superior to that derived from the weather balloon program, raised hopes for improved forecasts. *Weatherwise* kept meteorologists informed of technological improvements as the Weather Bureau acquired its own satellite, *Tiros I*, on April 1, 1960, equipped with two television cameras which could send back aerial photographs of widespread storms. Within the

[12] McKelvey, *Snowstorms and Snow fighting*, pp. 19–20; *Rochester Democrat and Chronicle*, Mar. 29, 1959; *New York Times*, Feb. 7 and 8, 1959; *Weatherwise* (Apr. 1959): p. 51.

[13] *New York Times*, Oct. 2, 1959; *Rochester Times Union*, Dec. 29, 30 and 31, 1959; David Ludlum, "Some Noteworthy Snows of 1959–60," *Weatherwise* (Feb. 1960): pp. 24–29. *Rocky Mountain News*, Sep. 30, 1959; *Denver Post*, Oct. 1, 1959.

year, *Tiros II* would carry still more accurate instruments aloft to scan the heavens and report on changing climatological conditions. It would take time to develop the full potential of the new information, as Jerome Namias, chief of the extended forecast section of the Weather Bureau had warned in 1957, but the prospect for more accurate forecasting was exciting to snow-fighters as well as meteorologists. New York City had benefited in December 1959, when a forewarning had enabled its sanitation department to get men and equipment out in time to clear the streets of a seven-inch storm and help residents really enjoy their white Christmas.[14]

Responsible snow-fighting officials were now checking the new weather predictions. When a widespread blizzard hit the Midwest early in February 1960, even hard-hit Milwaukee, was able to cope effectively with it. Des Moines, walloped by a two foot New Year's Day storm in 1942, when city and trolley-company plows had battled for three days to open the streets, had added salt and new plows to its arsenal; now, by restricting parking to one side of the streets and by operating its plows around the clock, it was able to open double lanes in most streets before the close of the second day. Omaha, which had benefited from the enterprise of Omaha Steel Works, now one of the major producers of snow-fighting equipment, got its plows, salt and cinder spreaders, and tow trucks out when the snow began to fall on February 9 and, despite some traffic snarls during rush hours that evening, had the streets open the next morning. But as the storms persisted, the battle to keep roads open became tougher. When the heaviest snow of the season hit on March 14th, three-

[14] *New York Times*, Jan. 5, Feb. 18, Dec. 23, 1959; Apr. 2, Nov. 24, 1960; Jerome Namias, "Weather Forecasting in Transition: A Survey and Outlook," *Weatherwise* (Aug. 1957): pp. 119–20, 140–42; "Tiros I Concludes Mission; Tiros II Readied," *Weatherwise* (Aug. 1960): pp. 159–60, 180; Thomas Malone, "1960: A Year of Excitement and Progress," *Weatherwise* (Feb. 1961): pp. 3–7.

fourths of Omaha's snow-fighting equipment was already in need of repair. With snow shovels out of stock, people bought grain scoops as substitutes. Motorists frantically sought tire chains to keep their cars operating. The city had spent $300,000 more than was budgeted for snow removal, but the job was essential, the mayor declared, as he dipped into funds set aside for highway improvements. When Omaha's "long winter" finally ended on the 16th, with twenty-seven inches on the ground and the total for the season at 56.1, the highest since 1912, the *World-Herald* welcomed a "Breather."[15]

Heavy storms generally required not only adequate preparation, but also a readiness to deal with the emergency. Thus Kansas City, which suffered the major blow from the March 1960 storm, also its heaviest since 1912, had to close all schools for the second time that winter. With part of its snow-fighting equipment still in the repair shop as a result of the February storms, many streets became clogged by drifts and stalled cars. Fortunately a decision by five major firms to close early and to remain closed the next day reduced traffic enough to enable the city, helped by rising temperatures and leased plows, to clear most arteries by the third morning.[16] The February blizzard, fed by moist clouds from the Gulf, moved eastward and dumped 18 inches on Rochester on February 14, but its snow-fighters were ready with 153 pieces of equipment, and kept the streets open. Syracuse, on the other hand, again saw its streets blockaded, its schools closed, and many stores and factories forced to shut down, and the state had to close the central and western portions of the Thruway.

A delegation of Finns who visited Rochester in the midst

[15] *Des Moines Register*, Jan. 2 and 3, 1942; *Des Moines Tribune*, Feb. 10, 1960; *Omaha World-Herald*, Feb. 10, Mar. 14, 15 and 16, 1960.

[16] *Kansas City Star*, Mar. 3 and 16, 1960; *Kansas City Times*, Mar. 3 and 16, 1960; *Rochester Democrat and Chronicle*, Feb. 15 and 22, Mar. 5, 1960; *Syracuse Post-Standard*, Feb. 15 and 16, 1960; *Weatherwise* (Feb. 1961): pp. 27–28; *New York Times*, Feb. 11, 1960.

of the raging blizzard expressed astonishment. Accustomed to heavy snows at home, they were surprised at the clogged condition of many streets and especially at the number of cars marooned in the drifts. Helsinki would "get the snow off the streets much faster," they declared, neglecting to add that the few private cars would be safely stored in stables. News arrived several months later of the rapid clearing of Moscow's streets after a heavy snow with the aid of six hundred scoop-plows, 1,700 dump trucks, and 4,000 shovelers. The *New York Times* correspondent who reported that feat also took note of the freedom from interfering traffic enjoyed in the Russian capital.[17]

As the winter storms moved eastward in 1960, both New York City and Boston reeled under the impact of their mounting fury. Taking advantage of an early warning broadcast, New York had its snow-fighting crews out and kept the streets open despite a 13.6-inch fall on March 3. But as the winds piled drifts into the entrances of the subways and the Hudson tunnel, traffic jams developed on the second day. Schools, libraries, and many offices were closed, and most flights to and from the airports were cancelled. But a spirit of camaraderie developed; even the taxi and bus drivers tried to serve and cheer distraught wayfarers. Boston, however, was more severely crippled. Hit by a 19.7-inch snow, the heaviest March storm on record, it quickly declared a snow emergency and suspended nonessential functions in a determined effort to clear its streets. Plunging temperatures considerably slowed the snow-removal operations and aggravated the plight of many residents. A reflective editor offered some consolation by observing that "a big storm is the folklore of the future."[18]

A bigger storm quickly obscures such memories, how-

---

[17] *Rochester Democrat and Chronicle*, Mar. 6, 1960; *New York Times*, Feb. 8, 1961, pp. 34–35.
[18] *New York Times*, Mar. 4, 1960, pp. 1–8; Mar. 5, 1960, pp. 8–9; *Weatherwise* (Aug. 1961): pp. 138–41, 167–70; Apr. 1969, pp. 74–75.

ever, and New York, along with Philadelphia, Washington, Buffalo, and Syracuse, would soon be absorbed by the harsher winter of 1960–61. An early storm, which dropped eighteen inches on Buffalo on November 30, proved to be only a forerunner of the widespread blizzard that buffeted a dozen northeastern states some ten days later. New York was coping successfully with a six-inch fall on the 12th as heavier blasts paralyzed its New Jersey neighbors, but as the snows continued, reaching 17 inches on the second day, the Big Apple, too, became crippled. Schools, airports, and many transit services ceased operations. Paul R. Severance, in charge of snow removal, mustered a force of nine thousand men for the biggest job in his twenty-four years as commissioner. And as New Yorkers looked back to the harsh December 1947 for a precedent, Philadelphia saw all records in twenty-five years surpassed. Both cities sought and secured National Guard assistance in rescuing stranded motorists and removing stalled cars from their highways.[19]

The winter, of course, had scarcely begun and showed no let-up. A six-inch snow that hit both New York and Washington on January 19, caused some traffic tie-ups in the metropolitan area and prompted the cancellation of several pre-inaugural events in the Capital. Although thousands of motorists were stranded in the drifts at Washington that night, a battery of snow-plows and sanders managed to clear the route of the inaugural parade by the next morning. A crowd of nearly a million people gathered to watch the procession of 30,000 marchers who saluted the newly-elected President John F. Kennedy under a clear sky that afternoon. As the storm continued unabated in New York, Mayor Wagner put the snow-removal forces on a twenty-four-hour schedule and authorized an expenditure of $5.4 million for the January clean-up. He reported an outlay in

---

[19] *New York Times*, Dec. 1, 1960, pp. 31–3; Dec. 12, 1960, pp. 1–2; Dec. 13, 1960, pp. 1, 34, 35 and 36; Local Climatological Data, Summary, Philadelphia.

the previous December of $6.9 million and warned of prospective additional costs. Comptroller Abraham Beame quickly went to Albany in quest of authority to float $15 million (increased to $20 million as the storm continued) in five-year notes to spread the emergency costs over (hopefully) normal years to follow.[20]

New Yorkers, accustomed to pressing ahead with their jobs whatever the obstacles, awoke with a start to the unprecedented crisis. As protests mounted, both from truckers unable to make deliveries on side streets, and from taxpayers appalled at the costs, a new storm on February 3 again closed the airports, highways, and schools and caused numerous power failures. Mayor Wagner declared a snow emergency and ordered the police to stop nonessential cars from entering the city. Transit company buses rescued over one thousand motorists trapped in marooned cars; welfare agencies supplied blankets to thousands of travelers stranded at the airports where 2,650 flights were cancelled. As the third big storm in six weeks slackened, after seventeen inches had fallen, a number of venturesome skiers appeared on Fifth Avenue; sleigh riders and snow-shoers flocked to Central Park, and merchants as well as householders began to clear their sidewalks. Schools re-opened on the 7th and, although snow-removal operations were incomplete, Wagner, after an extended inspection of conditions by helicopter, removed the emergency bans on the 10th, ending a full week's siege for which only the great Blizzard of 1888 supplied a fit precedent.[21]

---

[20] *New York Times*, Jan. 20, 1961, pp. 1–4; Jan. 22, 1961, pp. 1–2; Jan. 26, 1961; Jan. 27, 1961; Feb. 1, 1961; Hughes, *American Weather Stories*, p. 92.

[21] *New York Times*, Jan. 30, 1961; Feb. 4, 1961; Feb. 5, 1961, pp. 1–8, 44–6; Feb. 6, 1961, pp. 1–7, 16, 48; Feb. 8, 1961, pp. 1–6; Feb. 10, 1961, pp. 1–4. *Weatherwise*, Feb. 1961, p. 35; Apr. 1961, pp. 75–76.

Northeast Snow Conferences

The triple storms of 1960–61 would leave more than memories. The inspection by helicopter, the restrain of non-essential motorists, and a resort to bonds for snow-removal expenses were innovations of significance. Albany granted the bonding power, but the governor vetoed a legislative authorization of a ban on non-essential driving. The blizzards, however, had been widespread, and many cities had been plagued by unnecessary driving as well as by abandoned cars. Syracuse had had a thirty-six-hour snow emergency and Utica one that lasted two days.[22] Several northeastern cities had battled vigorously to re-open their streets, and as they tallied the costs after the crisis subsided, many were appalled. When New York City issued a call for a conference on snow removal, seven cities whose total costs had been more than $30 million responded.[23]

New York City, which had spent $22,700,000 on snow removal that winter, took the lead in pressing for an improved strategy. Frank J. Lucia, its commissioner of sanitation, invited suggestions from the attending delegates. Commissioners and superintendents from Washington, Baltimore, Philadelphia, Pittsburgh, Buffalo and Boston, reported on their own problems and experience. It soon became evident that more was needed than an exchange of questions and frustrations, and the conference decided to sponsor a research project on Urban Snow Removal and Ice Control to be conducted by the Research Foundation of the American Public Works Association. All agreed to gather again for a second conference a year later in Washington to launch the project.

The second Northeast Conference on Urban Snow

---

[22] *New York Times*, Feb. 19, 1961, pp. 72–79; Mar. 16, 1961, pp. 23–28; Mar. 23, 1961, pp. 21–25; *Rochester Democrat and Chronicle*, Feb. 7, 1961.

[23] *New York Times*, Mar. 1, 1961, pp. 63–67; Mar. 24, 1961, pp. 33–35.

Removal proved much more fruitful. Most of the former delegates were back after a year's reflection on their problems, and several were ready with positive proposals. Commissioner Lucia was convinced that the primary need was for effective snow-emergency traffic regulations, and he cited New York's action that year in listing certain streets as "snow streets" from which motorists had to remove all cars during a snow emergency. Commissioner Haley of Boston reported his adoption of a rule banning parking on the odd-numbered side of all streets to allow plows to push the snow into a windrow on that side and maintain two-lane traffic in the center. Most of the delegates reported light snows that winter, but Buffalo had again suffered a heavy blow in December when a twenty-six-inch fall had clogged its streets. Its policy of banning all parking along bus routes from 1:30 to 7:00 A.M. had helped plowing, but the collection of fines and removal of cars had presented difficulties. Each city had its own method of determining and announcing snow emergencies, and all agreed that a standard code was needed and should be produced by the projected research project.[24]

The conference devoted its second session to a discussion of the need for regional coordination of snow-fighting forces. Samuel Baxter of Philadelphia described the central command headquarters established by his city. Equipped with a teletype setup, it was able to relay orders on no-parking and other emergency regulations to radio and TV stations and to a task force of inspectors on patrol in cars with two-way radios. Washington had a common center with private wire connections to the schools, the highway departments of neighboring municipalities, and related city departments. New York had an Emergency Committee to help coordinate snow-fighting functions and

[24] Donald M. Fairlie, ed., *Northeast Conference on Urban Snow Removal, Washington, 1962* (Chicago: Amer. Pub. Works Assoc., 1962), pp. 1–12.

a city-owned radio to handle its communications. Lucia told of his mayor's keen disappointment, when a legislative measure granting the city power to halt the influx of suburban cars and ban non-essential motor traffic from certain city streets, had been vetoed by the governor. Commissioner Xanten of Washington observed that only New York with its subways could consider banning motor traffic, but that the Capital had partly achieved that end by virtue of the cooperation of the federal departments, which dismissed their staff early during a storm and declared a holiday on the second day if traffic obstructions continued. The commissioner from Pittsburgh doubted his city's ability to persuade the steel mills to shut down even for a day.

In a third session on snow-removal operations, William Xanten described the efforts of his department to ensure better cooperation from the public. Washington simplified its plan from one for five different kinds of storms to one for three easily-identified kinds. Cooperation was especially essential, he had found, in a city where most drivers were inexperienced with snow hazards. The program's success, he declared, depended in "a prolific use of salt" at the start of most storms and a timely mustering, as the storm developed, of plows and other equipment, to which specific individuals were assigned in advance. Buffalo reported its reliance on a district plan with men assigned to a specified district whose captain had authority to hire additional shovelers and engage private contractors as needed. Boston announced a plan to appoint a full-time manager of snow-removal functions with authority to develop and train a municipal force recruited from various departments, and to contract for the use of private plows during an emergency. The Pittsburgh delegate doubted the possibility of engaging private contractors during a storm when factories and stores were bidding for their services.[25]

Several common problems received close attention. The

---

[25] Fairlie, *Northeast Conference*, pp. 14–32.

decision as to how much equipment and salt should be kept on hand depended, Lucia declared, on the street mileage considered essential for the maintenance of city functions. New York, he reported, had designated 1,800 miles to be kept open during a storm, with the remaining 3.800 to be cleared as quickly as possible. Philadelphia and Washington had also designated about 30% of their street mileage for early care, but relied on a shift of weather to open the rest. Several questions were referred for further study; how quickly should the snow removal operation be launched and how soon completed?; how much salt should be stockpiled and where?; how much training operators of snow-fighting equipment should have and how many hours they should work?; and whether abandoned and illegally parked cars should be fined or hauled away, and at what fees? A query from the Buffalo delegate as to the number who made use of private forecasters produced answers from New York, Boston, and Milwaukee and expressions of full confidence in the Weather Bureau from Washington, Pittsburgh, and Philadelphia.

All looked forward hopefully to the contributions to be made by the research project. Commissioner Lucia reported a successful experiment in New York with a portable snow melter, but looked to the research project for data on the utility of tire chains on varied surfaces, and on the amount of salt appropriate for different temperatures; he strongly urged the preparation of a standard code for traffic control under snow-emergency conditions. The delegate from Pittsburgh hoped the research project would produce a conceptual breakthrough that would open new scientific approaches to the problem. Buffalo, as Edward Jurewiscz put it, would settle for operational procedures promising a more effective attack on snow blockades.[26]

The winter of 1962–63 produced no record-breaking

---

[26] Fairlie, *Northeast Conference*, pp. 32–48; *New York Times*, Aug. 26, 1961, pp. 5–19.

storms[27], but levied heavy charges on several snow-conscious cities. In August 1962, New York bought seventeen new salt spreaders with snow-plow attachments and acquired several additional snow melters. As the season progressed, developing a number of threatening storms in below-freezing temperatures, the city's officials announced several snow alerts banning parking and unprepared cars from 159 streets and calling key personnel to their posts, from which they generally were shortly released as the storms blew over. With a total snowfall of only 16.3 inches, the city cut its snow costs to $2.2 million in a winter rated by the Weather Bureau as one of the coldest in a hundred years.[28] Slightly warmer Philadelphia actually achieved that record but handled its twenty inches of light snow expeditiously.[29]

The succession of raging blizzards in the late fifties and early sixties, coupled with the more adequate coverage by national press services, radio and TV broadcasts, had stirred speculation concerning the onset of a new ice age. City-dwellers throughout the snow country, plagued by the hazards of even minor storms, were alarmed by reports of hardships suffered by a score of cities in four of the last six years. Occasional recollections of old-fashioned winters spaced ten or even twenty years apart with the sound of sleigh bells jingling through the crisp air, provided a nostalgic though false backdrop for the more sobering problems of a highly mechanized society. The fact that only a few cities in the New York snowbelts had suffered more than one severe blow in the current succession of

---

[27] Watertown, N.Y. was perhaps an exception. The 105 inches of snow it received in December and January exceeded that of any two consecutive months, but no single storm or monthly or seasonal records were broken. A. Boyd Pack, "The Heavy Snows of Watertown, N.Y.," *Weatherwise* (April 1963): pp. 66–67, 78.

[28] *New York Times*, Feb. 1, 1963; Mar. 16, 1963, pp. 4–5; Apr. 7, 1963, pp. 85–88.

[29] Ludlum, *Weather Record Book*, p. 54; Local Climatological Data, Summary, Philadelphia.

Snow-Fighting in a Metropolitan Era

hard winters (in other words, that the intensity of their impact was more in the medium than in nature's message) was not generally recognized. As a result, responsible administrators in many cities had an easier task than ever before in rallying support for their snow-fighting programs.

With more justification than most cities, Rochester, too, was *en garde*. New tests had revealed some toxic effects from the rust-inhibitor previously employed there and elsewhere, and its use was discontinued in 1962–63. Yet although its total snowfall of 76.4 inches that winter was almost ten below the seasonal average, the city spread a record amount of salt per inch of snow, totaling 34,260 tons at a cost of $341,299. It kept its streets clear and the sidewalks plowed, but some residents proved over-zealous in shoveling their driveways, as five deaths from heart attacks during a six-inch snow on January 24 demonstrated. The sudden rise in the annual toll of three or four such fatalities prompted the city health department to renew its warning to "shovel snow, not graves."[30]

The cold winter stimulated a revival of skating and other winter sports. Rochester, for example, kept eight natural skating rinks clear of snow and saw the opening of a third ski resort in its region and the addition of a chair lift at an older center. Buffalo, Syracuse, Utica, and Albany were more abundantly supplied with ski resorts that attracted skiers from a greater distance and, with the aid of newly invented snow-making machines, vied with the thriving resorts in New England. Ice boating revived on the Hudson River as well as upstate, and the first snowmobiles imported from Canada made their appearance. Wintertime sports were, of course, much more firmly entrenched north of the border, in the Laurentians near Montreal, and especially at Quebec where the annual winter carnivals had become a tourist attraction,

[30] *Rochester Times-Union*, Jan. 25, Apr. 8, 1963.

awakening visitors from the States to new delights in snow encounters.[31]

But the problems confronting snow-fighters were not forgotten. The third Northeast Conference on Urban Snow Removal drew delegates from fifteen cities to Boston in April. It again stressed the need for advance planning, flexible budgeting, and alert action when a storm started. The delegates discussed and approved an interim report of the two-year research project in which sixty-three cities had agreed to participate. A question about the efficacy of salt prompted *American City* to publish an extended review of a Highway Research Board Bulletin on that subject. Its charts specified the quantities of salt needed to melt ice and snow at varied temperatures, and warned against its use, except in generous quantities, when the thermometer dropped below 20° or 25°.[32]

A second cold winter in 1963–64 would spread down in the Midwest to give Memphis a record low of −13° on December 24 and a depth of fourteen inches of snow. It was sufficient to delay all traffic for a few days and to kindle reminiscences of earlier snowstorms in 1892 and 1948. The cold wave reached south to give New Orleans an astonishing four inches of snow on New Years Day.[33] As the storm drifted eastward, New York City received ten to twelve inches of snow, trapping thousands of cars in the streets and highways. Again the schools were closed and twelve deaths were attributed to the storm. The city put two thousand pieces of snow-fighting equipment into the fray, but it lacked sufficient tow trucks to remove the

---

[31] *Rochester Times-Union*, Feb. 26, 1962; Jan. 25, 1963; *Rochester Democrat and Chronicle*, Dec. 7, 1975; Floyd King, "Making Snow", *Rochester Democrat and Chronicle, Upstate*, Feb. 26, 1978; K. Revis, "Winter Brings Carnival Time to Quebec", *National Geographic Magazine* (Jan. 1958): pp. 69–96.

[32] *American City*, Jun. 1963, p. 162; Aug. 1963, p. 32.

[33] *Memphis Press Scimitar*, Dec. 23, 1963; *New York Times*, Jan. 2, 1964, pp. 22–29.

Snow-Fighting in a Metropolitan Era

abandoned cars, and the frustrated mayor appealed to suburban motorists to use the public transit facilities and leave their cars at home. He declared a snow emergency and banned parking on snow streets. Washington, with nine inches of snow, put its second snow-fighting plan into effect. It had suffered along with New York from the lack of an early warning of the storm's proportions, and the commissioners of both cities rejoiced when, at the fourth annual Northeast Conference on Urban Snow Removal, the Weather Bureau promised to improve its forecast warnings with the assistance of a computer able to process more quickly the data received from the weather satellites already in orbit.[34]

The fourth conference met at Philadelphia in April 1964, and discussed some of the findings of the research project soon to be released. All delegates were keenly interested in the comments of James F. O'Connor of the extended forecast branch of the Weather Bureau. But his promise of an increasing use of a high-speed computer in analyzing satellite photographs was tempered by his admission that this and other sources of data still left local weather predictions "in the realm of possibilities." A tardy or inaccurate forecast usually made the local weatherman the "fall-guy," as Xanten of Washington put it, but that status was soon shared by the mayor who failed to keep the streets open. No mayor, city manager, or commissioner could, he warned, meet the challenge of a heavy storm without advance planning and careful coordination of participating departments and allied agencies. A spokesman for the American Automobile Association noted the additional need for pre-storm training of area motorists in safe driving practices and, during a storm, repeated instructional broadcasts detailing alerts, traffic bans, specific hazards, and possible havens for those in trouble. A

---

[34] *New York Times*, Jan. 14, 1964, pp. 1–8; Jan. 15, 1964, pp. 1–18; Jan. 20, 1964, pp. 82–87; *American City*, Jun., 1964, pp. 106–07.

preliminary report of a study of traffic accidents under varied snow conditions in Chicago stressed the heavy costs involved, and stirred interest in a further study of their relation to the city's snow-fighting efforts.[35]

The research report, when finally released in August 1965, commanded close study in many of the sixty-seven cities and six states that had sponsored and funded it. Two chapters applying system-analysis to measure the cost and impact of varied depths and degrees of snow, and to the economies of snow and ice removal, displayed the technical skill of the research teams as well as the need for more explicit data on storm conditions. The other chapters emphasized the "many faceted" character of the snow problem; the potentialities, imperfect techniques, and uncertain costs of induced melting schemes; the urgent need for a responsible coordination of separate and often rival jurisdictions; and the imperative necessity of preparing an effective and feasible plan for combatting probable storms. None of these approaches was new, and none achieved a conceptual or technical breakthrough. But a chapter discussing snow ordinances supplied suggestive models for cities of varied sizes and situations. And the report in its entirety added a new scholarly dimension to snow-fighting. It provided "the basis for a broad, rational analysis of snow emergency activities," as Samuel Baxter, head of the Research Foundation, put it in his forward, but it would be up to each city to make that analysis and take the proper action.[36]

---

[35] *American City*, Jun. 1964, pp. 106–07; Robert K. Lockwood, ed., *Snow Removal and Ice Control in Urban Areas*, (Chicago: APWA Research Foundation 1965), pp. 3, 19, 21–22, 119. *See also*, Julius F. Bosen, "Data Processing Machines in Climatology", *Weatherwise* (Aug. 1962): pp. 158–59, 163; Morris Neiburger, "Meteorology in 1963", *Weatherwise* (Feb. 1964): pp. 3–7.

[36] Neiburger, "Meteorology in 1963", pp. ix ff.

Snow-Fighting in a Metropolitan Era

New Snow-Fighting Strategies

If the research report produced no dramatic solutions, its preparation did encourage a number of practical developments. The *American City* published an article tabulating snow-fighting equipment needs for heavy, medium, and light snows based on the population, street mileage, and temperature range. New York City developed a Master Snow Fighting Plan, which among other improvements designated four hundred street miles as snow routes to be cleared of cars as soon as a snow emergency was announced. Advance planning and pre-season training of personnel were now recognized as essential in several cities, and the no-parking ban on "snow streets", bus routes, and other crucial areas gained wider acceptance. When a new storm dropped five to nine inches on the city in January 1965, Mayor Wagner directed the police to stop cars lacking tire chains form using any of the snow-plow routes. A stiffer fine of $25 helped to increase compliance.[37]

The winter of 1964–65 brought additional cities into the snow-fighting arena. When Williamsport, Pennsylvania, was snow-bound by a heavy storm on the first of February, and Rome, New York, was paralyzed by a fall of 12 inches a week later, their states sent plows to help break open the principal streets.[38] Indianapolis battled a foot of snow from the same storm, and Detroit suffered a harsher blow at the end of the month. Not only its schools and libraries, but its auto plants and other factories were forced to close, as the motor capital struggled to free thousands of marooned cars and break its snow-choked bottlenecks. It was little consolation to learn that the number of new cars produced

---

[37] *American City*, Aug. 1964, pp. 75–77; Dec. 1964, pp. 87–88; *New York Times*, Jan. 11, 1965, pp. 3–6, 28–34.
[38] *New York Times*, Feb. 2, 1965, 2; Feb. 10, 1965, 2–3.

week was reduced from 301,725 to 155,333, a drop denied the city a new production record.³⁹

It was not a shortage, but a super-abundance of cars that plagued American cities when the storms blew in again a year later. Buffalo received the first severe blow, an eighteen-inch storm on January 23, 1966, the heaviest for that month in its long record. By good fortune it was a weekend storm, catching most residents at home, but many cars parked at the curb were effectively trapped. The city brought out its first contingent of plows, but soon discovered that in the recent shift of administrations forty-three trucks purchased for part-time use as plows had not yet been equipped with plow blades. Unable to break the drifts in front of some fire stations, the city declared a snow emergency. Most churches were closed that day and all schools the next day as Mayor Sedita negotiated a loan of plows from nearby Tonawanda. Buffalo had barely opened double lanes in most of its principal streets when a second storm hit eight days later. A weekday storm, it caught many residents at work in factories and stores, and traffic jams as well as chain-reaction pile-ups on the expressways complicated the situation. The schools were closed again, as well as the airport, the Thruway, and several key arteries. The city battled to re-open its bus and major traffic routes with the aid of a rotary plow mounted on caterpillar treads and a giant lift shovel to load a fleet of garbage.⁴⁰

That second storm was widespread, and Rochester was bombarded with two feet of snow during the next two days. The storm struck with unexpected intensity on Sunday evening while the author was addressing an annual banquet of the German-American Society in the ballroom of the city's leading hotel. On skipping out at the end of his review of their contributions to the city's history, he drove

---

³⁹ *New York Times*, Feb. 27, 1965, pp. 31–32; Ludlum, *Weather Record Book*, p. 75.

⁴⁰ *Buffalo Courier Express*, Jan. 24, 25 and 31, 1966.

from the city's newly-opened underground garage into a raging snowstorm. He was fortunately able to follow a plow to within a block of his house, but had to abandon his car in a drift at that point and founder the rest of the way through snow already almost a foot deep. Another plodder, blinded by the blowing snows, stopped him to ask directions to the nearest service or police station. On reaching home, he switched on the radio for further details, and learned that the several hundred German-Americans he had left dancing blithely in the hotel ballroom had learned too late of the storm's impassible dimensions, and had decided to continue dancing into the wee hours of the morning before bunking down in dress suits and ruffly gowns on extra mattresses in the ballroom. Some, unable to get home the next day, would resume dancing the second evening.

Of course no schools opened on Monday morning and most offices and factories quickly shut down. Many Rochesterians enjoyed a two-day holiday. The *Democrat and Chronicle* managed to get out a 36-page edition on Monday but could not deliver the 30,000 copies it printed. Fortunately the radio supplied a steady flow of news of storm closings and cancelled events as well as no-driving pleas and other directives. The city had deployed its snow-fighting equipment in advance, and made an early attempt to keep the streets open; it soon discovered, however, that the size of many drifts was too great for most of its plows. City Manager Scher leased bulldozers and power shovels from construction firms to clear clogged intersections and re-open vital lanes. And yet again, the most difficult task was the removal of abandoned or illegally parked cars. Even with some lanes opened, there were not enough public and private tow trucks to handle the load, and storage space was inadequate. The author got his car out of the drift on Tuesday afternoon, but joined the crowds waiting for buses that began to reappear the next day.

Several old snow hazards reappeared in new forms. Many people suffered heart attacks as they struggled to clear their driveways or free their stalled cars from the drifts. With supermarkets closed or inaccessible, householders had to seek supplies at neighborhood stores. Some merchants transported food and other essentials to hospitals by toboggan sleds, and the police used a snowmobile for emergency deliveries. When City Manage Scher, after long hours in the command post, removed the snow-emergency bans on the third day, he urged industrial employers to call in only essential workers, and opened the large Midtown parking garage without charge to attract cars off the streets. His report a month later put the cost of the three-day storm at $821,466, boosting the year's outlay on snow removal to almost $2 million; the business losses were estimated at $20 million.[41]

Meanwhile to the east, Syracuse and Oswego were more heavily blanketed with thirty-one and sixty inches of snow respectively. Oswego accepted its hardships with resignation, but Syracuse battled to maintain essential services. Thus Dr. Edward Hughes, a leading obstetrician, returning to the city in the midst of the storm, was dropped off at a suburban station when his train became stalled in the drifts. A call to the police for possible assistance brought a snowmobile to carry him, not to his home, but to the hospital where during the next three days he delivered babies to several mothers brought in by their frantic husbands. Funerals could be postponed as in former storms, but the happier rites of birth would not wait, and Dr. Hughes's experience must have been duplicated in part in many other snow-bound city hospitals.

As the frigid blasts of that northwesterner moved eastward, combining with a coastal storm, they dropped thirteen inches of snow in New York City and virtually shut down Baltimore and Washington. New York announced a snow emergency but lifted it within a day as its snow

---

[1] *Rochester Democrat and Chronicle,* Jan. 31, Feb. 1, 2, 3, 4 and 5, Apr. 12, 1966; *Rochester Times-Union,* Feb. 1, 2 and 5, 1966.

Snow-Fighting in a Metropolitan Era

fighters proved able to keep traffic moving. It was ready on the second day to send a convoy of four large rotary plows up the Thruway to help release motorists stranded along its course, thus enabling the state to despatch thirteen plows to Syracuse, while several counties came to Oswego's assistance. An army helicopter helped to locate stranded cars and buses, scattered dots on the white landscape, reassuring them of early release. The radio kept the public informed of developments and responsible to the needs of stranded victims. Otherwise, the technological advances of the "motor age" seemed only to have increased the storm's impact.[1]

---

[42] *New York Times,* Jan. 31, 1966; Feb. 1, 1966, pp. 1–3; Feb. 2, 1966, p. 20; Feb. 4, 1966, pp. 33–34; *Weatherwise* (Dec. 1966): pp. 240–47.

ROCHESTER'S ICE CONTROL FLEET, 1968
(Courtesy of Division of Public Information,
City of Rochester, N.Y.)

The proper use of salt was the most controversial issue in snow-fighting strategy during the 1950s and 1960s, and Rochester, with the Restof salt mine nearby, made a generous use of salt, tested the merits of additives, and finally reached a compromise with local environmentalists.

CHAPTER SEVEN

URBAN SNOW PROBLEMS WIDELY SHARED

If the technological advances of the motor age were complicating rather than mitigating the urban snow problems, they were also contributing, together with current demographic trends, to a transformation of the American urban-metropolitan landscape in ways that would promote new responses to the recurrent onslaughts of winter. The major cities in the snow country that had long borne the burden of snowstorm hazards had become central cores of sprawling metropolitan districts. Their original dominance over their statistical districts had begun to slip in the 1960s as city residents migrated to the suburbs which, by the late 80s would out-weigh their declining central cities both in population and wealth three or four to one. Despite repeated efforts, only one of the twenty-five cities in our snowfall table achieved metropolitan polity, and the response to snowstorms was also diffused. The 20 or 30% drop in population of the central cities did not reduce traffic in their streets, for the number of cars was mounting to two per family, and the increasing populations of their metropolitan hinterlands further swelled the motorized streams.

No wonder the chief hazard cited by the experienced snow-fighters at their annual conferences was not a heavy snowfall, but the parked or abandoned cars, whether on city streets or suburban highways. Clearly, neither plows nor salt would solve that problem. Authoritative regulations, adequate planning, alert and coordinated execution supplied the keynotes for successive North American Snow Conferences. Fortunately, under that broader title, the annual snow conferences attracted new delegates not only from Midwestern, Canadian and even some Sunbelt cities, but also from state and county highway departments, as well as federal climatological experts, thus assuring a wider sharing of snowstorm experiences and responses.

Formulating New Strategies

The central cities, despite their reduced size, continued to bear the major snow-fighting burden. Scattered across the snow country from Portland, Maine, to Fairbanks, Alaska, the twenty-five cities in our table, with mean annual snowfalls ranging upward from 20.9 for Philadelphia to 113.9 for Syracuse, would have ample occasion to test and revise earlier strategies as repeated snowstorms swept over their districts during the last quarter century. In fact, twelve of the twenty-five would see their earlier monthly records shattered and thirteen would suffer new seasonal highs, challenging their strategies. Several small cities in the snow country would record annual means higher than the Syracuse maximum, and a few would try ingenous new tactics. A dozen borderline cities with moderate annual means would also face one or more heavy blasts, posing challenging questions to meteorologists. One conclusion all would reach was that no city stands alone in facing the ravages of winter.

All of our major snow country cities had formulated plans and accumulated snow-fighting equipment considered appropriate for their size and climate. The fears of an approaching ice age, prompted by the successive harsh winters in the 1950s, would be muted and forgotten as atmospheric scientists discovered gaps in the Earth's ozone layer and warned of a more catastrophic warming trend. A

Urban Snow Problems Widely Shared

more immediate challenge to our central cities was to find resources and enlist responsible allies in confronting the snowy onslaughts that repeatedly deluged their burgeoning metropolitan districts.

Milwaukee and Chicago both confronted that problem in January 1967 when a record-breaking storm inundated the west coast of Lake Michigan. Milwaukee, a central city that still dominated its metropolitan district, received 59.5 inches that season, but having suffered heavier blows in previous years, was prepared to cope with such storms. Wisconsin, used to harsh winters, had state plows to clear essential highways to the city line and kept the city's links with its suburbs open.[1] But Chicago, hit by successive blizzards in January and February that totaled over 50 inches (68.4 for the season), saw all records shattered and was "staggered" by the repeated blows. Looting in the deserted Loop of stores and stranded cars resulted in 188 arrests and the shooting of two suspects who refused to stop. With a snow-fighting arsenal valued at $8.3 million, Chicago had barely opened traffic in its principal arteries, clogged by a twenty-one-inch fall on January 26, when another storm dropped eleven inches on February 2. The clean-up operation, compounded by clogged highways in the suburbs, continued for another two weeks and included shipments of snow by trainload to Tennessee in search of adequate dumps. Costs soared to $8 million and the business losses to an estimated $100 million, and Mayor Daley ordered a new study of the city's snow-fighting needs.[2]

The report by Barton-Ashman Associates, submitted a year later, cautioned against an attempt to assure clear streets in all storms. Such a program would require an

---

[1] *New York Times*, Jan. 27, 1967, pp. 22–26; Local Climatological Data, Annual Summary, Milwaukee, 1935–1975.

[2] *New York Times*, Jan. 27, 1967, pp. 22–26; Feb. 3, 1967, pp. 16–17; *Weatherwise*, Apr. 1967, pp. 68–70; Dec. 1967, pp. 248–53.

annual expenditure of $10 million on maintenance and stand-by services, with most of it unused nine years out of ten. The city and transit authority should be prepared, the consultants declared, to handle a nine-inch storm promptly with publicly-owned equipment, but Chicago should negotiate agreements in advance to rent additional plows, loaders and other machines from private contractors as needed in heavy storms. They further recommended a tighter control and closer coordination with suburban snow-fighting forces, a speedier call-up of workers, and improved parking bans among other traffic controls. They estimated that delays in salting and plowing during rush-hour traffic in January had added $100,000 to the city's cost, a needless waste.[3]

New York City, with a century of experience in snow-fighting and with metropolitan control throughout its teeming boroughs, was accustomed to battling even minor storms. Yet the relief it had enjoyed during earlier storms from its elevateds and subways had been eroded as the more intense commercial, residential, and industrial developments these facilities promoted had drawn trucks, taxies and private cars in from the expanded city, clogging many streets and highways at all hours from dawn to past midnight, in fair weather as well as foul. Thus when a storm hit, salt spreaders often showered more pellets on stalled or parked cars than on the road surface and plows had to proceed at a slow and deliberate speed. Even a three-inch snow generally called for an announcement of a snow emergency. When the tail-end of Chicago's second blizzard reached New York on February 7, dropping 12.6 inches on Manhattan and much more in other boroughs, the city faced a real challenge.[4]

One problem in the city, as the *Times* noted, was that "all New Yorkers hasten to their jobs when a storm starts.... Many leave early and stay home on the second day, but the

[3] *Public Works*, Aug. 1969, pp. 86–90.
[4] *New York Times*, Feb. 8, 1967, pp. 1–5, 14–15.

crush in the streets becomes incessant." Thus on the 7th, many who had driven into the city found their cars marooned, and Mayor Lindsay broadcast an appeal to motorists to leave their cars in parking lots and go home by public transit. All schools and many institutions, including theaters, closed that day, as well as the airports where many travelers remained stranded. Fortunately the snow-fighting forces coordinated by teletype dispatches to their division heads were able to open many streets by the second day, while radio and TV announcements kept the public informed of their progress. A recruitment of additional workers on the third day released most of the sanitation men who had manned the dump trucks, freeing them to tackle the mounting piles of garbage, which frequently compounded New York's snowstorm ordeals. The appearance of boisterous youngsters with sleds on suburban slopes engendered a holiday spirit that marked the end of the siege for some; many motorists, however, still had to dig their cars from the huge furrows that lined streets and highways. The city faced a snow-removal bill of $9,400,000 for the season.[5]

The same storm spread up the coast from Washington to Portland, battering many communities. Washington, with ten inches on the ground, had been considerably relieved as the federal offices declared a holiday, letting the city tackle its snow-removal task as on a weekend. Boston was more seriously strangled by an eight-inch snow that trapped many cars and halted all planes. Portland, with a record-breaking 45.8 inches that month, more than any of its fellow sufferers, was accustomed to winter sieges and took its time digging out. The *Press Herald*, in fact, was more attentive to the storm's impact on winter sports than to the blockade it created in the streets.[6] A city of modest

---

[5] *New York Times*, Feb. 8, 1967, pp. 1–8; Feb. 9, June 22, 1967;
[6] *New York Times*, Feb. 8, 1967, p. 15; *Public Works*, Aug. 1967, pp. 96–98; Local Climatological Data, *Annual Summary*, Portland, 1935–75; *Portland Press-Herald*, Feb. 8, 1967.

size (65,000 in 1970), Portland merits inclusion in our table because of its frequent bouts with east-coast storms and its relaxed handling often of their severest blows. Thus, two years later when it took the brunt of still another storm, which raised its snowfall total for February 1969 to a record-shattering 61.2 inches, the city brought out its plows, loaders, salters, and blowers, and managed to keep some traffic moving throughout the three-day storm, winning more plaudits than criticism. It was a strategy other small and a few larger cities would learn to emulate.[7]

The storms of 1967 and 1969 stirred an interest in snow-fight practices in Canada and elsewhere abroad. An article from Montreal, a city that rivaled the New York snowbelt cities in annual snowfall, struck a happy note as it told of the efforts of the provincial highway department to keep the toll road into the Laurentians open for skiers and other winter sportsmen from the city. At the toll stations, electric heating coils in the pads assured their constant use. Another article told of the expenditure of $300,000 by Ottawa in one storm, and of $1.75 million during that eighty-inch snow season, in a determined effort to keep the capital open throughout its centennial year. Canada as a whole was spending $100 million annually on snow removal, the article reported, to maintain service on its railroads, highways, and airports as well as to clear city streets, and was able to welcome skiers and winter visitors from near and far.[8] Several cities in northern Europe were equally ebullient in responding to snow. Stockholm, for example, spent approximately $2 million annually for snow removal and kept its routes to the ski slopes open. Few cities on the Continent suffered heavy snowstorms, but many now had plows and salt spreaders ready for

[7] *New York Times*, Feb. 27, 1969; *Portland Press-Herald*, Feb. 27 and 28, 1969; *Weatherwise*, Apr. 1969, pp. 69–72; Dec. 1969, pp. 230–35.
[8] *Public Works*, Feb. 1967, pp. 103–105; *New York Times*, Jan. 21, 1968.

action, for an increasing number of motorists were demanding bare streets there, too. Only in Moscow could two inventors blithely promote an improved roller to hardpack roads for sleighs and cars.[9]

Most American cities were taking a sober cost-accounting view of their snow problems. Mayor Lindsay, appalled at the cost of the 1966–67 season, investigated and found evidence of kickbacks paid by the major snow-hauling contractor to an acting superintendent. He promptly fired that official and canceled the contract. Fortunately a mild winter enabled him to reorganize the department with strict economies in mind without injury that year.[10]

During the two relatively harmless winters in the snowbelt region, interest focused on a pair of experiments conducted under a snow-modification program sponsored by the federal government. The object, to test cloud-seeding techniques as a possible means of diverting snows approaching Buffalo to the ski slopes some sixty miles to the south, proved partially successful but raised such protests from residents in the "deposit area" that the experiments were suspended. A later test at Buffalo explored the possibility that the city itself, by its profuse discharge of smoke and other industrial effluents, was contributing to its own snow records; however the stratagem of relocating the factories, many of them on the city's south side where the snowfalls were frequently very heavy, was not considered.[11]

It was not industrial smoke, but a combination of metropolitan pressures that escalated a moderately heavy snowstorm in New York City in 1969 into an urban blizzard. If the wind velocity was less than violent, the public furor made up for it. A misleading prediction by the Weather Bureau, forecasting an early end of a moderate storm, made Mayor Lindsay delay calling his emergency council

---

[9] *New York Times*, Mar. 9, 1968, pp. 41–44; Feb. 16, 1969, p. 63.
[10] *New York Times*, Dec. 28, 1967, pp. 1–3; Dec. 29, 1967, pp. 1–14.
[11] *New York Times*, May 14, 1968, pp. 1–49; Feb. 24, 1970, pp. 45–48.

into session. His commitment to economy had permitted division heads, relying confidently on salt, to neglect the upkeep of little-used equipment. As the snow deepened and many sanitation workers, unable to reach their assigned posts, reported to more accessible stations, the department sent 1,800 of them home until a protest by their union and the urgency of the storm forced the mayor to reverse the order. Thus the storm acquired the upper hand before a snow emergency was put into effect.[12]

The great city was caught in a trap of its own making. Its schools closed promptly, but the airports, refusing to announce cancellations, drew a host of scheduled passengers, six thousand of whom became stranded in the waiting lounges. Many commuters who tried to drive to work had to abandon their cars in the streets. The outward migration of Manhattan workers had vastly increased the number of such drivers and had dispersed the staff of vital departments. When the snow-fighters finally reached their stations, they found an estimated 40% of the machines out of order or unequipped with plow blades, some of them buried in the snow in the parking lots. By hard work, the sanitation workers began to open the streets of Manhattan and Brooklyn on the third day, but residents of Queens, still bogged in fifteen inches of snow, protested loudly. Trying to show his concern by striding through the drifts in Queens, the mayor attracted cat calls rather than cheers. Governor Rockefeller declared the metropolis, excepting Manhattan, a disaster area and provided assistance in recruiting 10,000 workers to help clear the streets and restore order.[13]

James Reston penned a satirical column chiding the city for making Lindsay the scapegoat, but he did not exculpate the mayor who was never again mentioned as a presidential

---

[12] *New York Times*, Feb. 10, 1969, pp. 1–8; Feb. 11, 1969, pp. 1–7.

[13] *New York Times*, Feb. 11, 1969, pp. 41–46; Feb. 12, 1969, pp. 26–29; Feb. 13, 1969; Feb. 17, 1969, pp. 1–2; Feb. 18, 1969, pp. 40–41.

possibility. Other critics cited the success of snow-fighters in Milwaukee and Buffalo in keeping their streets open. Rochester, too, supplied an innovate lead when, plagued by a traffic jam during a three-inch snow early in November that winter, the city manager had amended its snow-emergency plan by instituting a "pre-snow-emergency" alert to call police, tow trucks, and key public-works supervisors into action when snow began to accumulate.[14] As a result, Rochester maintained its services in a relatively mild winter (79.7 inches, more than double New York's 30), and Lindsay, observing the merits of the snow alerts there, hastily called one on March 2 when a light snow, predicted to reach six inches, started. That storm quickly subsided in New York City, but it was a close call as twelve inches fell on several of its New Jersey neighbors. The mayor quickly placed orders for new equipment that summer. He held an inspection of his forces and the new equipment on December 22, 1969, and three days later, when a storm hit Christmas morning, the sanitation department sounded an alert that called fifteen hundred plows and two hundred salt spreaders into action. Lindsay, who had flown to the Bahamas for the holiday, rushed back to direct the work. "The only way to beat a storm is to get the jump on it," he declared as he assigned additional men to some of the fifty-eight district forces.[15]

It was Albany that took the brunt of that storm. Hit by an intensive fall on December 26, capping a record-breaking month that totaled 57.5 inches, Albany was paralyzed. With all traffic halted, its officials and relief agencies had to speed emergency supplies by snowmobiles. The state had reverted to its earlier policy of subsidizing county road departments and had no plows to spare, but fortunately

---

[14] *New York Times*, Feb. 11, 1969, pp. 1–31; Feb. 14, 1969, pp. 3–38; *Rochester Times-Union*, Nov. 15, 1968.

[15] *New York Times*, Mar. 2, 1969, pp. 62–64; Mar. 3, 1969, pp. 31–35; Dec. 26, 1969, pp. 1–3; Dec. 27, 1969, pp. 4–10.

Mayor Corning's plea brought a contingent of plows by Thruway from Niagara and Erie Counties to help break the drifts on the third day.[16]

The bruising character of that storm, demonstrating the need for more adequate equipment, prompted Governor Rockefeller to recommend an appropriation of $4 million for snow removal to enable the state to assist stricken communities. Mayor Lindsay requested funds to buy seventy-five bombardier plows to speed the clearing of sidewalks. It was a cold winter with freezing temperatures extending from the middle of December through January, and now to the mayor and his snow-fighting officials every approaching cloud posed a threat. The city sounded twenty-five snow alerts, calling out many workers; it recorded eleven snowstorms, all of them minor, but if the total snow was five inches below average and readily handled with salt, the strain on the force was great, and the department described the winter of 1969–70 as the worst in forty years. Perhaps to substantiate that report, the winter, two days later, produced a late March storm that canceled the Easter parade.[17]

It had not been a light winter upstate—quite the reverse. Albany had enjoyed a respite after its December storms, but Buffalo, Rochester, and especially Syracuse and Oswego, all in the Erie and Ontario lake-effect snowbelts, had taken a pounding. Neither Oswego's 191-inch winter total, nor Syracuse's 125.5 inches matched earlier snowfalls, but the 51.9 inches that deluged Syracuse in December broke all previous records for that month and inflicted grievous hardships on its residents. Yet articles on snow in Syracuse generally devoted less attention to the progress of its plows than to the prospects of its skiers, and its snowmobilers enjoyed a bonanza that winter. Buffalo and Rochester

[16] *New York Times*, Dec. 30, 1969, pp. 1–8; Dec. 31, 1969, pp. 33–35.
[17] *New York Times*, Jan. 14, 1970, pp. 32–37; Feb. 1, 1970, pp. 53–54; Mar. 27, 1970, pp. 39–41; Mar. 30, 1970, pp. 1–3.

declared snow emergencies and tackled the job of clearing their streets with spirit. Last minute Christmas shoppers were somewhat reduced in Rochester during an eleven-inch storm on December 24, and air travel was delayed at times in both cities, but Buffalo managed to open its major highways on the second day after a twenty-three-inch storm.[18]

Buffalo got a partial break the next winter but Oswego, Syracuse, and Rochester sustained still heavier blows. In fact, the total snowfall at Syracuse, 157.2 inches, shattered all previous records, while Rochester's 142.7 took second place in its fifty-year totals. City plows, salt spreaders, and snow-loaders tackled the job in each city and, with the aid of snow-emergency regulations and rented lift plows and dump trucks, snow-fighters managed to keep the major roads open. The Thruway was closed on two occasions, and the airports as well as the schools were shut down for varied periods, but the cities experienced less disruption than each had suffered in some previous winters. Rochester officials rejoiced as one fifteen-inch snow hit late on Friday night, giving the snow-fighting crews a weekend to clear the largely empty streets. By Monday they were able to pronounce all main streets open. The snow-removal outlays soared to new heights at both Rochester and Syracuse and climbed again in the succeeding winter, which proved only slightly less severe. But again the journalists at Syracuse gave greater attention to the conditions on neighboring ski slopes than to the city's problems. Business at the ski resorts was brisk, and snowmobiles were a hot item there and in Rochester, too, where the demand for skis, high-top boots, and related items gave evidence that many residents, as one reporter put it, "love a good snow storm."

---

[18] *Public Works*, Aug. 1970, pp. 58–60; *Rochester Times-Union*, Dec. 27, 1969; *Syracuse Post-Standard*, Dec. 26, 1969; Local Climatological Data, Annual Summaries, Buffalo, Rochester and Syracuse; *Weatherwise* (Dec. 1972): p. 277.

Snow in the Cities

Unfortunately the number of heart attacks and other snow-related fatalities was also high.[19]

Among the major cities, only Montreal in Canada suffered comparable blows. Indeed its 1970–71 total of 165 inches established a new metropolitan high for the continent and its city engineer sent a report on "Facing Snow Emergencies" to the *Public Works Journal*. His biggest problem had not been the snow, which he was prepared to salt and plow with adequate provincial equipment, but the private cars parked or stranded in the streets. Even a new scheme, which equipped plows and spreaders with two-toned horns to announce their approach, failed to clear the streets, many of which remained impassable for thirty hours early in March. The only movement for a time was by snowmobiles, and the city decided to buy a number of such machines for future use when the snow would again have to be measured in feet rather than inches.[20]

No other major cities rivalled the snowfall records of Montreal or the three snowbelt cities, but a number of smaller cities topped even their maximums, Juneau in Alaska, for example. Long silent sufferers of snowstorms, small cities and towns were attracting new interest in their responses. John F. Rooney, an urban geographer, studied the snow hazards of seven midwestern and mountain cities. By measuring the disruptions suffered in transportation, trade, manufacturing, and other fields, and relating these to the depth, moisture content, and wind velocity of specific storms, he made an appraisal of their relevant hazards and responses. Only the most hard-pressed—Milwaukee, because of its size, and Muskegon on the eastern shore of Lake Michigan—had developed well organized snow-fighting

---

[19] *New York Times*, Dec. 18, 1970; Jan. 28, 1971; *Rochester Times-Union*, Jan. 14, Feb. 15 and Mar. 5, 1971; Feb. 21, 23 and Nov. 15, 1972; *Rochester Democrat & Chronicle*, Dec. 14, 1970; Feb. 15, March 5, 1971; Feb. 5 and 22, 1972; *Syracuse Herald-American*, Mar. 3, 1974.

[20] *Public Works*, Aug. 1971, pp. 79–82; *Weatherwise* (Dec. 1970): pp. 270–72.

programs. But if the public outlays elsewhere in his sample were moderate, the private costs Rooney tabulated were considerable. Residents of Rapid City, Casper and Cheyenne, his three western towns, spent, he estimated, more than $600,000 annually on tire chains in their private bout with winter storms. The search for a balance between private and public costs and potential injuries was, in Rooney's opinion, the crucial task facing urban officials in the snow country.[21]

Muskegon's alertness to its storms in the snowbelt region east of Lake Michigan was characteristic of other lake-effect snowbelt towns and small cities—Oswego, Watertown, and Dunkirk in New York, Kalamazoo and Grand Rapids in Michigan. Each had early snow-fighting plans but relied in part on state help. Oswego, one of the most frequently victimized, dramatically demonstrated its hazards before the annual meeting of the North American Snow Conference in February 1972. None of the hundred or so delegates, who arrived without difficulty despite the visible accumulation of two earlier storms, expected the snow burst of fifty inches in twenty-four hours which held them snowbound for four days. Only the arrival of a contingent of state plows helped break the drifts that topped twenty-six feet in some districts. It was a new and memorable experience for most of the delegates.[22]

When the Conference met in Cincinnati nine years later, its superintendent of highway maintenance reported on a survey of the snow-fighting strategies and expenditures of the fifty American and twelve Canadian cities that had responded to his questionnaire. Grouping them by size and weight of snowfall, he found that while small cities had generally responded vigorously to light storms, most of them had turned hopefully to the state for relief from heavy falls; the large cities, brushing off the light falls, had

---

[21] John F. Rooney, Jr., "The Urban Snow Hazards in the U.S." *Geographic Review* (Jan. 1967): pp. 538–59.
[22] *Weatherwise* (Dec. 1972): pp. 276–87.

exerted themselves against the heavier snowfalls and vividly remembered their battles.[23]

By the 1970s, most state highway plows no longer stopped at the city line, a policy shift that helped small cities more than most large ones, though their suburbs benefitted greatly. Small as well as large cities were concerned over the cost of snow-fighting, and a few boldly experimented with technological strategies. Thus Brookline, "The Bay State's Largest Town," rented a snow-melter in 1967 and claimed it spent far less during that blustery season. Some large cities acquired melters, notably New York, but they remained a minor part of the snow-fighting arsenal.[24]

Even some of the borderline cities, scattered westward from Washington to Seattle, with low annual snowfall means that enabled them to rely chiefly on salt, occasionally suffered a heavy blow that provoked a strategic innovation. When Seattle took the brunt of a blustery northwest storm in January 1969, its transit system put an emergency snow-route schedule into operation, enabling the city's plows and salt spreaders to clear alternate streets and keep traffic moving. The first snow blast coming on Sunday had helped that strategy, but as the storm continued, reaching a depth of nineteen inches, the city had to ask the National Guard for helicopter supervision and snowcart assistance in breaking snow blockades and motor tie-ups in the suburbs. Fortunately, Lockheed and two other shipping yards closed for two days, reducing traffic, and by equipping its buses with a fresh set of tire chains each day, the "Queen City," with "an armament of red and yellow scrapers, brushers and sanders," kept street traffic moving and the airport runways open throughout the four-day siege.[25] Tracing a wide arc, another erratic storm dropped a heavy

---

[23] *American City*, Aug. 1981, p. 17.
[24] *Public Works*, Aug. 1967, pp. 96–98.
[25] *Seattle Post Intelligence*, Jan. 27, 29 and 30, 1969.

snow on Kansas City, which responded efficiently with plows and salters; unabated, the storm continued south and dumped 16.5 inches on unprepared Macon, Georgia, where it trapped an estimated 25,000 cars on city and state roads, plaguing the astonished officials.[26]

Most cities in the snow country had, by the mid-seventies, equipped themselves with plows, spreaders and tow trucks to handle possible ten-inch storms. The Chicago strategy had taken hold in the midwest where several of the heavy storms of these years raged, establishing new records. Denver, which received a near record-breaking total of 94.9 inches in 1972–73, Omaha, which took the brunt of a widespread January blizzard in 1975, and Detroit, which suffered an accumulation of 35 inches in December 1974, each tackled its drifts with an armament of city plows, spreaders and trucks, but quickly rented additional bulldozers and tow cars when necessary. Manufacturers in Omaha and other machine-tool centers were designing new and speedier plows and more efficient gauges for salt spreaders; tire manufacturers were producing snow tires of such quality that the demand for tire chains was down, but their practice of implanting metal studs to increase the grip on icy roads was proving destructive to highway surfaces and strict regulations as to their use were enforced.[27]

Mounting Opposition to Salt

Concern over the use of salt was escalating still more rapidly. Motorists, eager to use the streets and highways throughout the year, had become resigned to springtime repairs of rusty cars, even those protected by the manufacturers with anti-corrosion coatings. Many citizens, regarding such

[26] *Weatherwise* (Oct. 1974): pp. 192–96.
[27] *Public Works*, Jan. 1971; Aug. 1972, pp. 60–61; Aug. 1973, pp. 74–75, 94–95. Dec. 1974, pp. 74–75; *Omaha World Herald*, Jan. 11, 13 and 19, 1975.

damages a lesser hazard than skidding accidents, supported city and town administrations that maintained dry pavements. Environmentalists, however, questioned the accident-prevention claims of heavy salt users and submitted conflicting though inconclusive statistics of their own. In the late sixties, when public health investigators found excessive quantities of chlorides in some public wells, as well as salt contamination in ponds and other bodies of water, a chorus of protests arose. Horticulturalists cited injuries to roadside trees, especially sugar maples and other highway foliage. Sportsmen deplored the depletion of fish in salt-polluted streams, and reported an increasing number of deer and other animals killed as they licked the drains of salted highways. Highway engineers complained of the effects of salt on road surfaces and particularly on the structure of bridges, some of which had to be rebuilt or resurfaced every year.[28]

But it was the threatened contamination of ground water and the imperiled safety of drinking water that brought the most positive action. Minnesota, where frigid temperatures reduced the usefulness of salt, took the lead in banning its use except on hills, intersections, and high-speed highways. Legislators in Massachusetts proposed a similar ban, and the town of Burlington in that state, alarmed by a 283–parts-per-million sodium chloride reading in its public wells, high above the 250–parts danger level set by the United States Public Health Service, banned the use of highway salt altogether. Environmentalists in several cities launched studies of the effects of de-icing salts on ground and drinking water and on plant life, and called for corrective action.[29]

---

[28] *Public Works*, Aug. 1971, pp. 64–66; *American City*, Sep. 1970, pp. 81–85; *Highway Research Record*, Nov. 11, 1963; *Park Maintenance*, vol. 18, no. 2.

[29] *New York Times*, Apr. 22, 1973, pp. 31–37, Mass. Legislative Research Council, *Report Relative to the Use and Effects of Highway De-icing Salts*, (Boston, 1965).

Responsible officials could no longer disregard the challenge to their leading snow-fighting weapon. A paper on "Highway Chlorides—Menace or Manna," delivered at the North American Snow Conference at Chicago in April 1971, was summarized in *Public Works*. The author reviewed the various hazards complained of, but noted that a study conducted by the Highway Research Board of the Public Works Association had blamed part of the salt found in ground water on salt deposited in normal rainfalls. Since that deposit could not, however, explain the increased salt pollution of recent years, he stressed the damage caused by the improper storage of salt and by its excessive application on bridges to keep them from freezing. He concluded that precautions were in order, but that salt remained indispensable to snow-fighters.[30]

William E. Dickinson, president of the Salt Institute, endorsed that conclusion. The Institute's "Sensible Salting" program called for the proper storage of salt in covered piles on water-proof pads guarded by drainage ditches; it also recommended the use of improved spreaders equipped with gauges able to control the quantity and direction of the salt stream to assure the greatest benefit. In low temperatures, a mixture of calcium chloride with salt was proposed to speed action, but its high cost discouraged extensive use and left the cheaper sodium chloride as the only practical de-icing agent. For airports, worried whether salt would threaten the structural safety of expensive aircraft, several alternative chemicals were suggested, including urea, which was an effective de-icer and free of corroding aspects. The Buffalo airport, among others, had substituted urea, despite its high cost, for salt to help its speed-plows, sweepers, and blowers clear the runways.[31]

---

[30] *Public Works*, Aug. 1971, pp. 64–66; Highway Research Board, *Environmental Considerations in the Use of De-icing Chemicals*, (Highway Research Record, 193, 1967).
[31] *Public Works*, Dec. 1971, pp. 54–56, 57–62.

As the debate continued, an intense search for alternatives to salt developed. A study sponsored by the U.S. Environmental Protection Agency was abstracted in *Public Works* under the title, "Blue Sky Approach to Snow and Ice Control Alternatives," It noted the merits of urea and other de-icing chemicals, and of pavements heated by steam pipes or electric wires, but concluded that the costs, ten to thirty times that for salt, restricted their use to limited facilities such as runways, ramps, and key driveways. Efforts to develop a coating for road surfaces that would prevent ice from making a hard grip had proved disappointing,[32] as had efforts to use electro-magnetic energy to crack ice for easier plowing, but some new rubber-tipped plow-blades attracted interest. When the 13th North American Snow Conference, held in New York City in April 1973, focused its attention on the salt question, angry biologists, foresters, and other environmentalists received a hearing, as did the Salt Institute and its allies, but again the debate was inconclusive.[33]

Some environmentalists were ready to settle for greater precautions in the use of salt. Discouraged by the limited effectiveness of sand, Massachusetts returned to the use of salt mixed with calcium chloride for certain purposes and scattered by a more carefully controlled spreader. Burlington, after reducing the salt content in its water supply, built a new reservoir and returned to a limited use of salt; but on the advice of the Massachusetts officials, it quickly provided covered storage and adequate drainage for its

---

[32] In 1982–83, Allentown, Pa. would apply a coating of Verglimit, a patented de-icing product developed in Germany, to the streets in a four-block area, and was able to report better driving conditions there, with one plowing after a heavy snow, than on other streets handled with salt, sand and frequent plowing. Whether its effect would last for the ten years needed to recoup the cost remained to be seen. APWA *Reporter*, 1983, pp. 36–37.

[33] *Public Works*, Aug. 1973, *New York Times*, Apr. 22, 1973, pp. 31–37.

stockpiles and also limited the amount of salt used to 2,000 tons a year.[34] In Rochester, sometimes called the "Salt Capital of America" because of its generous use of that agent, a Committee for Scientific Information, organized to fight pollution in its varied forms, sponsored a number of studies designed to promote improved methods of storing salt and greater precautions in its application. By avoiding a blanket condemnation of salt, the committee won a hearing for its analysis of the wasteful usage and reckless storage practices in some suburban towns as well as in the city, and promoted several precautionary reforms. Its findings, supported by a regional group of the Sierra Club, led the city to cut is salt use by 30% over the next two years and helped to persuade the county to shift to a 3 to 1 mixture of sand and salt. A careful tabulation of the number of traffic accidents in the city and county under the new salting program during the winter of 1974–75 showed no increases.[35]

A Testing and Sharing of Strategies

The increased awareness of the hazards of salt spurred renewed efforts in many cities to improve other aspects of their snow-fighting programs. As it happened, a pair of "Stand-out Winters" hit the North American snow country in 1976–77 and 1977–78, testing established practices and

---

[34] *Public Works*, Jan. 1974, pp. 50–51; Aug. 1974, pp. 50–52.

[35] Lindsay K. Holmes, "Environmental Effects of De-icing Salts," (Dec. 1973); ibid., "Salt Storage in Monroe County," (Dec. 1973); ibid., "The Use of De-icing Salt in Monroe County" (June 1974); ibid., "Accidents and Salting in Monroe County" (Feb. 1976). All the preceding typescripts were prepared for the Rochester Committee for Scientific Information; Laurie Hindson, "Salt City: USA" *Rochester Democrat and Chronicle*, and *Upstate*, Mar. 24, 1974.

moting a wider sharing of urban snow-fighting techniques.

Snow holidays and other wintertime pauses acquired a more austere quality in these harsh winters. The relaxed attitude even of some snowbelt cities during the relatively moderate winters of the mid-seventies disappeared as a record-breaking cold wave spread down from Canada in January 1977, plunging temperatures throughout the snow country and far into the South to unusual lows. Winds across the lakes battered Buffalo and its suburbs with a succession of blizzards, clogging streets and highways with depths that surpassed the city's previous winter snowfall record of 126.4 inches by mid-January. The onslaught forced the announcement of snow emergencies and snow holidays to make plowing easier, but while the well-equipped and experienced snow-fighting forces of Buffalo and Erie County managed to re-open major roads after each storm, permitting a relaxation of restraints, renewed blizzards piled up new drifts, trapping thousands of cars. The unrelenting siege finally boosted the city's snow total by the end of the month to a new metropolitan high for the nation. When the season ended in April, it had reached 199.6 inches.[36]

Buffalo, the first city to secure federal aid in battling a snow blockade forty years earlier, again received emergency assistance from National Guard units and from a detachment of three hundred army engineers rushed to the scene in late January. And after the fourth blizzard hit, again paralyzing the area, President Carter, responding to the pleas of local and state officials, finally declared the snow-clogged six-county district a disaster area and granted federal disaster assistance, the first time in snowstorm history.[37]

---

[36] *Buffalo Evening News,* Jan. 13, 15 and 31, Feb. 1, Apr. 16, 1977; *New York Times,* Jan. 14, Feb. 1, 1977; Ludlum, *American Weather Book,* p. 16.

[2] *Buffalo Evening News,* Jan. 31, Feb. 1, 1977; *New York Times,* Feb. 2 and 6, 1977.

Buffalo's ordeal set other precedents as well. Mayor Stanley M. Makowski, on discovering that his announcement of a "snow emergency" did not provide adequate restraints on drivers, proclaimed a "state of emergency" to enable police to enforce a ban of all unnecessary driving. When the snow-fighting crews re-opened some major highways, he relaxed the ban to permit the operation of cars carrying three or more passengers. His appeals for outside assistance brought not only state and federal aid but also a detachment of eight rotary snow-plows by airlift from New York City. The discovery of several frozen bodies in cars buried in drifts, and reports of increased looting and of an upsurge in burglaries, recalled earlier experiences in Chicago and elsewhere during snow blockades, but Buffalo, with over 200,000 idle workers, as plants and offices remained closed into a second week, faced a crisis situation, and the mayor had to secure federal assistance in the form of an emergency food stamp program to ease the hardships of many families.[38]

Buffalo, now the undisputed "Snow Capital", was not the only sufferer in that harsh winter. Its frigid winds carried freezing temperatures deep into the South, blighting fruit and vegetable crops in Florida, and covering Chesapeake Bay with a coat of ice disastrous to its fisheries. As the bitter cold spread over the eastern two-thirds of the nation, creating new demands for heat, it precipitated a natural gas shortage of crisis proportions.[39] The efforts of the Carter Administration to meet these challenges by declaring disaster areas in Florida and Maryland, and by promoting conservation of fuel and other measures are subjects for a separate study. But surely the increasing national awareness of these related emergencies helped to

---

[38] *Buffalo Evening News*, Feb. 1 and 5, 1977; *New York Times*, Feb. 2 and 6, 1977.

[39] *New York Times*, Feb. 1, 2 and 6, 1977; *Rochester Democrat and Chronicle*, Feb. 1, 1977.

induce federal action on the snow-fighting front. Thus in addition to the six counties in the Buffalo area granted national catastrophe aid, President Carter extended similar assistance to three counties in northern New York State where snows off Lake Ontario dumped huge deposits on Watertown and its neighbors as the winter progressed. National Guard units equipped with bulldozers and other snow-fighting machines joined the battle against snow drifts and removed stalled cars in several other snow-battered cities. Heavy though not generally record-breaking snows, aggravated by the rigors of the fuel shortage, made 1976–77 a harsh landmark winter throughout the snow country.[40]

Few weather-watchers, accustomed to sharp climatological fluctuations, anticipated the onset of a second severe winter. Some, with long memories, could have recalled pairs of back-to-back harsh seasons in the snow country, but none could have predicted a second landmark winter that would surpass its predecessor in the number, extent, and severity of its blizzards. Indeed, except in their meteorological extremes, the two winters were not very similar. The prevailing storm track of 1977–78 followed a wider arc than its predecessor and carried successive cold waves across the southern fringe of the snow country from December through February. St. Louis, Louisville and Cincinnati, as well as Columbus and Indianapolis, which generally escaped such blows, each suffered blizzards that left record-breaking levels of snow. And as the cold waves fanned out to the east, transforming moisture-laden streams from the south into raging snowstorms, Washington, Philadelphia and New York, and most cities farther north, buckled under snowfalls that challenged earlier records.[41]

---

[40] *New York Times*, Feb. 6, 1977.
[41] *New York Times*, Jan. 28, Feb. 8, 9 and 28, 1978; *Local Climatological Data*, 1988, for Columbus, Indianapolis, Philadelphia and Washington.

Improved Weather Bureau services, assisted by satellite photographs of potential storms "stacked up" over the north Pacific, enabled meteorological broadcasters to keep millions of viewers informed of impending onslaughts. Most people, however, after shaking their heads over early-morning reports of distant ravages, quickly drove to work or shop ahead of approaching storms, and in many cases became trapped before nightfall. Superintendents of snow-fighting forces, some warned in advance by private forecasters, frequently saw their crews and equipment impeded by traffic snarls. Even snow-emergency regulations, generally announced after a severe storm had developed, served only to mitigate the crisis, and several states and most stricken cities moved to invoke the sterner state-of-emergency curbs on traffic and unnecessary movement, in order to give their snow-removal forces a chance.[42]

Indeed, after battling with indifferent success against four successive storms in December and January, the hard-pressed officials in Rochester immediately declared a snow holiday when a new blizzard, billed as "the storm of the century," threatened on January 26. All schools were quickly closed, and factory and office workers were dismissed by early afternoon. Speeding home under threatenng skies but on dry roads, many motorists stocked up at neighborhood markets for a blizzard that completely bypassed the city. "We've never done anything like this before," declared the head of the Industrial Management Council, who estimated the business losses from the shutdown at $6 million. Yet almost everyone enjoyed the extra holiday, and the refuse crews were able to clear away some of the mounds left by earlier storms and get the city ready for the next blizzard, which struck with real vengeance on a Sunday evening ten days later. With a foot in the ground the next morning, there was no debate over the announcement of a snow holiday, and most motorists,

---

[42] *New York Times*, Feb. 8, 9 and 10, 1978.

having finally gotten the message, kept their cars off the streets, giving buses and emergency vehicles, including plows, the right of way. Buoyed by a more light-hearted spirit, many residents began to speculate on how soon the city would top its snow record of 161.7 inches.[43] Nearby Syracuse, buried under a heavier blanket by that blizzard, did not quite match its January record but registered a new high for the season, allowing it to reaffirm its status as "No. 1."[44]

The timing and other circumstances of the widespread blizzards of 1978 differed in each case, but everywhere the weight of the deluge and the height of the drifts trapped thousands of cars in the streets and on suburban highways and brought travel to a standstill. Even the more adequately equipped cities, such as Cleveland, hit by a lake-effect downpour of 42.8 inches, that gave it a new record for January, had to lease private bulldozers to clear the streets. Cities unaccustomed to heavy snows, such as Columbus and Providence, relied partly on state aid. Governor James A. Rhodes of Ohio, appalled by the number of fatalities caused by the "killer blizzard," declared a state of emergency and called out the National Guard to help rescue marooned motorists and to tow cars blocking streets and highways. As the storms and the havoc spread eastward, the governors of six other states invoked state-of-emergency regulations; in several cases, notably in Ohio, Connecticut and Massachusetts, the governors rushed state forces into the fray and made state funds available to enable hard-pressed communities, urban and suburban, to lease private equipment. Impressed by these efforts, President Carter declared Connecticut, Massachusetts and Rhode Island disaster areas.[45]

---

[43] *Rochester Democrat and Chronicle*, Dec. 27, 1977; Feb. 7, 1978.

[44] *Syracuse Post-Standard*, Dec. 27 and 28, 1977; Feb. 6, 8 and 12, 1978.

[45] *New York Times*, Feb. 8 and 9, 1978; *Local Climatological Data*, 1988, Cleveland and Columbus.

The January 1979 blizzard was, in effect, demonstrating the merits of the new strategy painfully worked out at Buffalo the year before. James L. Pierakos, the executive in charge of its snow-fighting operations, had addressed a session at the North American Snow Conference in Rochester on April 21. After briefly reporting the statistical measures of Buffalo's ordeal, and listing the nine neighboring and distant governmental agencies that had lent substantial assistance, he proposed a strategy of "Mutual Aid for Snow and Storm Disasters." It should include a sharing of municipal equipment and personnel, support by state and federal agencies, and authority to adopt and enforce clearly-defined state-of-emergency, snow-holiday and other regulations. Several of the cities pummeled by the 1979 blizzard had consciously or by chance followed parts of that strategy, though only one had declared a snow holiday, and none moved to negotiate agreements for mutual aid as he proposed.[46]

Only Rochester, encouraged by the assurance of Pierakos that a couple of additional holidays would have helped in Buffalo, had declared a snow holiday to reduce traffic in its streets. But in Washington, the birthplace of the holiday, the tactic had faltered. Hit by a record-breaking 2.5-foot snow, federal offices had closed for all except essential workers, but too many had considered themselves "essential", and driving into the Capital, had joined the flow of sightseers in lengthy tie-ups that blocked the snow-fighters and threatened chaos. Only the fortunate presence of a delegation of farmers, many of whom had driven in on their farm tractors in support of a bill (and were eager to demonstrate their good will by towing abandoned cars and breaking drifts) allowed Washington and its sprawling suburbs to resume normal activities on the third day.[47]

---

[46] *North American Snow Conference*, Rochester, Apr. 20–22, 1977; Letter of J. L. Pierakos to B. McKelvey, May 5, 1977.
[47] *New York Times*, Feb. 20, 1979; *U.S. News*, Mar. 5, 1979.

If the use of a snow holiday as a tactic in snow-fighting strategy was neglected, except for the widely accepted school holidays, the holiday spirit snowstorms engendered was spreading. Portland and Syracuse journalists were not alone in featuring the attractions of neighboring ski resorts when snowstorms hit, and sales of winter sports equipment, including the newly developed snow-making machines, were zooming in many cities, offsetting some of the losses their merchants reported after heavy storms.[48] Minneapolis and St. Paul, the coldest metropolis in the country, as Ludlum described it, hosted an "International Conference on the Livable Winter City," in March 1978. Under the leadership of Professor William C. Rogers of the University of Minnesota, it encouraged an acceptance of snowstorms as natural experiences and stressed the opportunities they presented for joyful activities. Among the innovative suggestions considered at the conference were the planting of evergreen trees to enhance the winter landscape, the clearing of skating rinks on ponds in the parks, snow sculpture contests, a re-introduction of the use of snow rollers instead of plows on selected streets, an increased construction of skyways downtown, the erection of colorful fishing shacks on nearby lakes, the charting of ski trails and, of course, ski runs on nearby slopes. Professor Rogers, Director of the Minneapolis Committee on Urban Environment and chief promoter of the conference, hoped to see Minneapolis become the nation's model Livable City by 2000.[49]

---

[48] An economist should make a critical study of the reported industrial and commercial losses suffered from major storms, for deferred production and deferred sales during a holiday or closure may not appreciably affect the annual returns, which depend on the market. Real losses would be suffered by workers laid of without pay, as Buffalo discovered.

[49] William C. Rogers, "The Beautiful Winter City", presented to the Minneapolis Committee on Urban Environment, 1978; International Conference, *The Livable Winter City*, Mar. 19–21, 1978, Spring Hill Center, Minneapolis; John Siegenhaven to B. McKelvey, Mar. 22, 1978.

Urban Snow Problems Widely Shared

Of course the holiday spirit seldom received attention at the annual meetings of the North American Snow Conference. Its advisory council, comprised in 1977 of seventeen representatives of competing producers or marketers of snow-fighting equipment, three of salt suppliers, and four editors of related journals, naturally focused the successive conferences on practical matters. Sessions presenting the capabilities of competing machines shared the three-day program with panel discussions of techniques for plowing, salting, dumping, traffic control, management, financing and the like. One additional subject of keen interest was the frequent reports by representatives of the Weather Bureau on current and planned improvements in its crucial services. Frequent articles in *Weatherwise* and elsewhere reported developments of the *Tiros* satellite program and other meteorological efforts to supply accurate warnings of weather prospects. Most sessions and articles on the subject concluded with a reminder that all weather reports beyond a day or two were still prognostications, not predictions.[50]

Snowfighters on the front line in the cities were able to benefit from the improved forecasts. No winter from 1979 to 1989 produced as harsh or as widespread havoc as the two standout winters discussed above, but some storms proved bruising enough. The February 1979 blizzard that gave Washington a monthly record of 30.6 inches was wide-spread. It dropped a record 27.6 inches on Philadelphia and 33.1 on Baltimore, well above their annual expectations, creating serious blockades in their streets and suburbs. Portland, Maine, had received and handled a record deluge of 62.4 inches in January with its accustomed calm.[51]

---

[50] *Weatherwise*, Oct. 1977; Jun. 1982, p. 228; April 1984; *Science*, June 1987, pp. 493–99.
[51] *Local Climatological Data* 1988 for Baltimore, Philadelphia and Pittsburgh.

Chicago, however, reeled under the repeated onslaughts of December and January that winter, and finally proved the weakness of its much-admired 1967 strategy. Mayor Daly quickly spent a million dollars upgrading his equipment by adding twelve new rotary-plows, but his failure to adopt stern snow regulations and permitting leading firms to cancel shifts *after* rather than *before* insufficient workers arrived, thus increasing the jams of slowly moving or stalled cars, compounded his problems. None of the successive storms quite equaled the two of 1967, but a simultaneous cold spell, plunging temperatures to $-14°$ one night in January and averaging 12.3° for the month, effectively sidelined the city's fleet of salt spreaders and prolonged the blizzard's grip. The extended siege gave Chicago a new monthly snow record in December 1978 and a new seasonal record of 83.7 inches for 1978–79.[52]

The major storms of the next three seasons swept through the Midwest. One in February 1980 hit in central Illinois, prompting the governor to declare a snow emergency in eighteen counties and send state plows to their rescue. Cleveland took the brunt of another storm in November 1981 which, with lake-effect increments in the succeeding months, boosted its seasonal total to a record 100.5 inches. Fortunately, state and suburban county plows helped keep the highways open.[53] Massachusetts would do the same for Boston, hit by an east-coast storm the next December and January, though the airport as well as schools had to be closed on two occasions.[54]

The central city's need for collaboration with its suburban towns and counties, as well as with the state, was gaining recognition. County and small-city delegates at the annual snow conference were weighing the merits of rival plows and salters and discussing emergency regulations

---

[52] "How Chicago Coped with the Blizzard," *Fortune*, Feb. 26, 1979.
[53] *New York Times*, Feb. 27, 1980.
[54] *New York Times*, Dec. 7, 1981; Jan. 12, 1982.

appropriate for their size and location. A few were initiating new tactics, among them, Shobie, Illinois, which made use of its air-raid system to alert residents by specified blasts of the emergency parking and driving rules to be enforced during a storm. Wisconsin prescribed rules for the selection of roads to be plowed during a storm, giving priority to those leading into city "snow streets". Superhighway authorities were assuming responsibility for clearing the snow from expressways leading into and through the central cities.[55]

By a partial application of the Buffalo strategy, Minneapolis and St. Paul, with county and state aid, withstood the rigors of three storms that broke their seasonal record in 1981–82; they topped it again in 1983–84 without losing their holiday spirit, as St. Paul demonstrated by staging its annual Winter Carnival.[56] An east-coast storm in February 1983 battered Baltimore, Philadelphia and New York City, yet despite traffic tie-ups in Manhaatten, the worst since 1974 as the *Times* reported in full detail, a spirit of camaraderie prevailed; skiers flocked to Central Park and a horse and sleigh appeared on Fifth Avenue with bells jingling.[57] Buffalo and its suburbs suffered a succession of storms for three years in a row, and in January 1985 three of its four district counties received state-of-emergency powers to combat their drifts. Early action and cooperation moderated the impact.[58]

Three erratic storms in the mid-eighties swept down from the Northwest into the deep South. The first, in January 1985, described by Ludlum as "The Storm of the Century," buried San Antonio under 13.2 inches of snow,

---

[55] *American City*, Feb. 27, Dec. 11, 1981.

[56] *Weatherwise* (Feb. 1983); *Local Climatological Data*, 1988, Minneapolis and St. Paul; *New York Times*, Feb. 1, 1986.

[57] *New York Times*, Feb. 13, 1983; *The New Yorker*, Apr. 26, 1985, pp. 27–28.

[58] *New York Times*, Jan. 12, 1982; Jan. 22, 1985; *Rochester Democrat and Chronicle*, Jan. 27, 1985.

paralyzing the city and stranding cars on area highways. Turning eastward, it delivered an unprecedented blow to Jackson, Mississippi. City and state highway departments shared unaccustomed tasks there and again in New Mexico in January and December 1987, when two additional storms plunged through the mountains to blanket El Paso and block traffic over a vast area, closing four airports. A sideswipe by that last storm hit Denver, burying some of its suburbs under a 42-inch blanket and blocking traffic throughout the district.[59] While these storms produced no new snow-fighting techniques, they raised questions among meteorologists at the National Center for Atmospheric Research which had drawn scholars at seventeen universities into a long-term research project know as GALE: The Genesis of Atlantic Lows Experiment. Another project further north, the Canadian Atlantic Storm Program, hoped to identify some of the factors controlling Gulf, East-coast and Northwest storms and increase the Weather Bureau's ability to track and time major storms. In hopeful anticipation of improvements to come, *Weatherwise* published an article in 1986 by Samuel Milner on "The Coming Revolution in Radar Storm Detection and Warning."[60]

The North American Snow Conference drew a record six hundred delegates to Des Moines in April 1988. Karen Fisher, a frequent contributor, gave a talk on "Creative Solutions to Snow Control" in which, among other innovations, she commended a move by Chicago to expand its use of X-ban Sear Radar to keep track of road conditions during a storm. A session of the previous year at Hartford on "Managing Snow-Control Budgets" was again under discussion at Des Moines, where another speaker stressed the merits of annual training programs for snow-fighters

---

[59] *New York Times*, Jan. 15, 1985 and Jan. 19, 1987; *The New Yorker*, Feb. 18, 1985; *Weatherwise* (Apr. 1985).

[60] *New York Times*, Dec. 24, 1985; *Weatherwise* (Apr. 1986).

and cited the offerings of a Snow School at Colorado S University.[61]

One of the traditional east-coast storms in the late 1980s would again prompt a federal snow holiday at Washington in 1987, drop fifty inches on Portland, and a record-breaking 46.8 inches on Worcester. Syracuse experienced four snowy seasons, totalling 426.8 inches (letting it retain its "No. 1" status, while Buffalo and Rochester welcomed a breather. Buffalo continued to stage its annual Blizzard Ball, and although New York had dropped that practice with the passing of the last survivors of its great blizzard of 1888, the centennial of that landmark event inspired a number of commemorative articles.[62]

Rochester's respite from heavy snowstorms in the late 1980s was terminated by a blistering ice storm in March 1991, which it would have traded for the worst blizzard, even that of February 1958, for there was little its plows, blowers and salt could do to relieve it. Fortunately, the havoc wrought among its canopy of trees won disaster-area support from the federal government, which assisted in a protracted campaign to remove the debris, topple the damaged trees and replant for a restoration of the Flower City's long cherished street and park foliage.[63]

In the early 1990s, despite climatological predictions of a warming trend threatened by gaps in the ozone layer over Antarctica, successive winters produced several widely-scattered blizzards. Minneapolis suffered a record-breaking storm in December 1991 but calmly maintained its "Livable Winter" programs. Salt Lake City received a lake-effect deluge from Great Salt Lake in January 1993 and its school children were given extra holidays. Worcester and Boston, with pre-training in the use of updated

---

[61] *American City*, Apr. 1987; pp. 27–32; Apr. 1988, pp. 56–65; Jun. 1988, p. 26; *Public Works*, Sep. 1988, p. 128.

[62] *Weatherwise* (Apr. 1988).

[63] *Rochester Democrat and Chronicle*, Mar. 4, 6 and 10, 1991.

equipment and increased aid from the state, survived, along with Syracuse, two heavy mid-winter storms, and when the "Great Blizzard" of March 14, 1993 hit them, all three established new seasonal highs.[64]

But that east-coast blizzard, frequently called a "White Hurricane", wrecked more havoc on the South, unprepared even for moderate snows. Striking, as predicted, Key West, it swept up the Florida peninsula and raced north in a wide swath across the coastal states, dropping one or two feet of snow on towns and cities, blocking highways, closing airports and schools and many factories, and causing power outages in many southern communities. It exacted a toll of ninety-two lives, mostly in the unprepared South. Baltimore, Washington, and Philadelphia, with some experience, were able to cope, benefitting from the providential timing of the storm, which hit on a Sunday. But even New York City, best-equipped and most experienced, found the storm's mixture of snow and sleet, with changing temperatures and velocities, a frustrating onslaught. While it quickly opened its highways, the mounds of snow and sheets of ice would plague the city for two weeks, and boost its cost to ten million dollars. It was, many commentators agreed, an historic, climatic storm, the greatest of the century in the millions afflicted and the area ravaged, demonstrating that Mother Nature had not slacked her power. But two comforting conclusions emerged: the Weather Bureau, by issuing a forecast four days in advance, accurately locating, timing and weighing the storm, had challenged Nature's inscrutability, and northeastern cities, having learned to take weather warnings seriously, had been able to prepare for the onslaught and took it in good stride, making full use of

---

[64] *Local Climatological Data*, Minneapolis, 1991; Salt Lake City, 1993; Worcester, 1993; Boston, 1994; *New York Times*, Feb. 12, 13 and 15; Mar. 4, 1994.

the emergency powers and assistance granted by their states.[65]

The wider sharing of responsibilities was promoted at successive North American Snow Conferences, which now attracted many county and state officials. As one speaker put it: "No city stands alone" in its wintertime ordeal. And as the meteorologists expanded their research to a global dimension, the plight of cities on the northern fringe of Europe and Asia came into view, as well as their surprising responses in the eighties and nineties.[66]

Abundantly supplied with snow by moisture-laden winds from the Sea of Japan to the west and the Pacific to the east, Sapporo, Japan's northermost commercial and industrial center, a city of over a million inhabitants and devoted to winter sports, had served as host of the Winter Olympics in 1972. But its continued growth had intensified some urban wintertime problems, and its mayor, Takeshi Itagaki, decided in 1982 that consultation with the heads of other northern cities would prove beneficial to all. His widely-scattered invitations attracted officials from nine cities to what became the first Northern Intercity Conference held in Sapporo's Cultural Hall in February 7-10, 1982. None of the cities that attended could rival Sapporo's snowfall records, which exceeded even those of the New York snowbelt, but among those present, Edmonton in Canada, Minneapolis in the States, and Helsinki in Finland, reported more frigid January temperatures, and the conference produced a score of useful proposals, several of which Mayor Itagaki found worthy of immediate application: time and place regulations for the use of studded tires, the planting of evergreen trees along streets and in the parks, an increased emphasis on (and variety in)

---

[65] *Rochester Democrat and Chronicle*, Mar. 14, 15 and 16, 1993; *New York Times*, Mar. 15 and 16, 1993.

[66] *North American Snow Conference*, Rochester, Apr. 1986; Des Moines, 1988; Hartford, Apr. 1989.

winter sports, the use of sodium lamps on athletic fields and other public areas, and the organization of citizen groups to participate in the care and full use of parks throughout the winter. "Heavy snows," he declared, "should be recognized as an asset, not a liability."[67]

The conference at Sapporo spurred the formation of a Winter City Association to promote the conduct and scheduling of similar conventions bi-annually in key northern cities, and Shenyang in China became the second host in September 1985, attracting ten participating cities including Chicago and Portland, Maine. Among its recommendations was the convening of a Winter Cities Forum to bring together scientists, city planners and officials involved in snow-fighting operations, to focus their attention on the problems of northern cities. Edmonton, designated as the third host city, staged the first Winter Intercity Forum in 1986, which proved so successful that it staged a second in conjunction with its Northern Intercity Conference in February 1988. It opened a Winter Cities Showcase, displaying products for wintertime use and enjoyment, and followed it with a Winter Expo to which 107 corporations submitted winter-relevant exhibits.[68]

Not content with the snowballing of its discussion meetings and displays, the WCA launched a number of research projects, one studying the environmental problems resulting from the use of salt in snow-fighting, another seeking and evaluating alternative antifreeze substances, and still another appraising the relative costs of the installation and operation of heated water pipes and electric wires in crucial street and sidewalk surfaces. It proposed and partially implemented a computerized Intercity Information Network to promote a continuous sharing of experiences and response, in order to fulfill the WCA's objective of "Living

[67] *Winter Cities Now*, (Sapporo, Japan, 1989).
[68] *Northern Intercity News*, Nov. 1989, p. 3; *Northern Intercity Third Conference*, Edmonton, 1988.

in harmony with winters." The launching of the *Northern Intercity News* in November 1989, a four-page semi-annual report, contributed to that objective.

Twenty cities sent delegates north to Tomaso, Norway, on the edge of the Arctic Circle for the fourth conference in March 1990, and thirty-four attended at Montreal in January 1992. A report of the research committee on The Winter Urban Environment, which had surveyed the snow-removal and road management practices of twenty-one cooperating northern cities, provided a major focus for the Montreal sessions. The discussions revealed a desire for more sophisticated guidelines, and the research committee re-tackled the snow-removal problem and came up with a *Technological Manual on Winter Road Management* for the sixth conference at Archagle in March 1994.[69] The WCA, reorganized as the International Association of Mayors in Northern Cities, was still headquartered in and run by Sapporo, a rapidly growing city where snow-removal problems commanded high priority in its plan to provide its now 1,700,000 inhabitants with a "Livable City" environment by 2000, and, to win recognition as a winter resort for an increasing flood of winter sportsmen and tourists.[70]

Thus the "evil snow," as George Templeton Strong characterized it in 1879, has somewhat lost its demonic qualities. Cities have not covered themselves with crystal domes, as he predicted, but several have built subways, airways and domed sports arenas, and all have covered shopping plazas and huge parking garages to take cars off the streets. New York City's pioneer efforts at snow removal in 1877 have been taken up and expanded in city after city, and they have finally, as we have seen, secured county, state and even federal assistance. The mystery of changing weather patterns has not been solved, but meteorologists are collecting enough information with sufficiently sensitive recording

---

[69] *Northern Intercity News*, Dec. 1990, p. 1; 1991, pp. 1–4.
[70] *Snow—Sapporo 21 Project*, (The City of Sapporo, Jun. 1991).

instruments to supply urban officials with timely warnings of storm probabilities. Technological advances have proved a mixed blessing, complicating the snow problems but increasing the city's capacity to combat them. Hazards persist, but as Buffalo demonstrated, after suffering the harshest siege of the century, they can be mastered by the responsible cooperation of all involved. And when a blizzard blows over, as it always does, it becomes, as one distraught mayor put it, "the folklore of the future."

But the city's experience of, and response to, snowstorms is more than folklore; both have been vital aspects of America's urban history. The original conception of harsh snows as expressions of Divine wrath against man's transgressions gave way as the growing towns discovered that heavy snows were obstructions that could be surmounted by horse-drawn sleds and sleighs, that eased transit and supplied wintertime entertainment for a full century. And as cities developed and acquired horse cars and railroads, they cleared the tracks of snow with horse-drawn or steam-driven V plows, which had to be replaced, after the arrival of trolley cars, by rotary-plows and dump carts to remove the mounds. Soon the festoons of wires that carried the city's power, light and messages were buried for safety from storms, a boon to the urban landscape. Again, as the gasoline motor cars replaced the trolleys, the city not only motorized its snow-fighting armament, but assumed full responsibility for clearing the streets, a significant development in urban polity. Meanwhile, salt, originally a minor aid in clearing railroad switches, became a major aid in combatting ice and assuring the dry roads motorists demanded, but its wide use and corroding and contaminating effects stirred nascent environmentalists into action. And as the cities developed into sprawling metropolises, the single-handed snow-fighting efforts of the declining central cities required and eventually received assistance from their mushrooming suburban towns, counties and states.

Urban Snow Problems Widely Shared

The American urban society is still in process of development. New mechanical inventions or new scientific discoveries may produce changes in the city's snow-fighting strategies as in other urban affairs. It is also possible that the newly activated cooperation in snow-fighting may contribute to the development of a more responsible metropolitan polity. Meanwhile, people with increased leisure time are already discovering the pleasures of a snowy landscape and the joys of winter sports, and northern cities can vie for "livable gold stars" to match their heroic "snowfall medals".

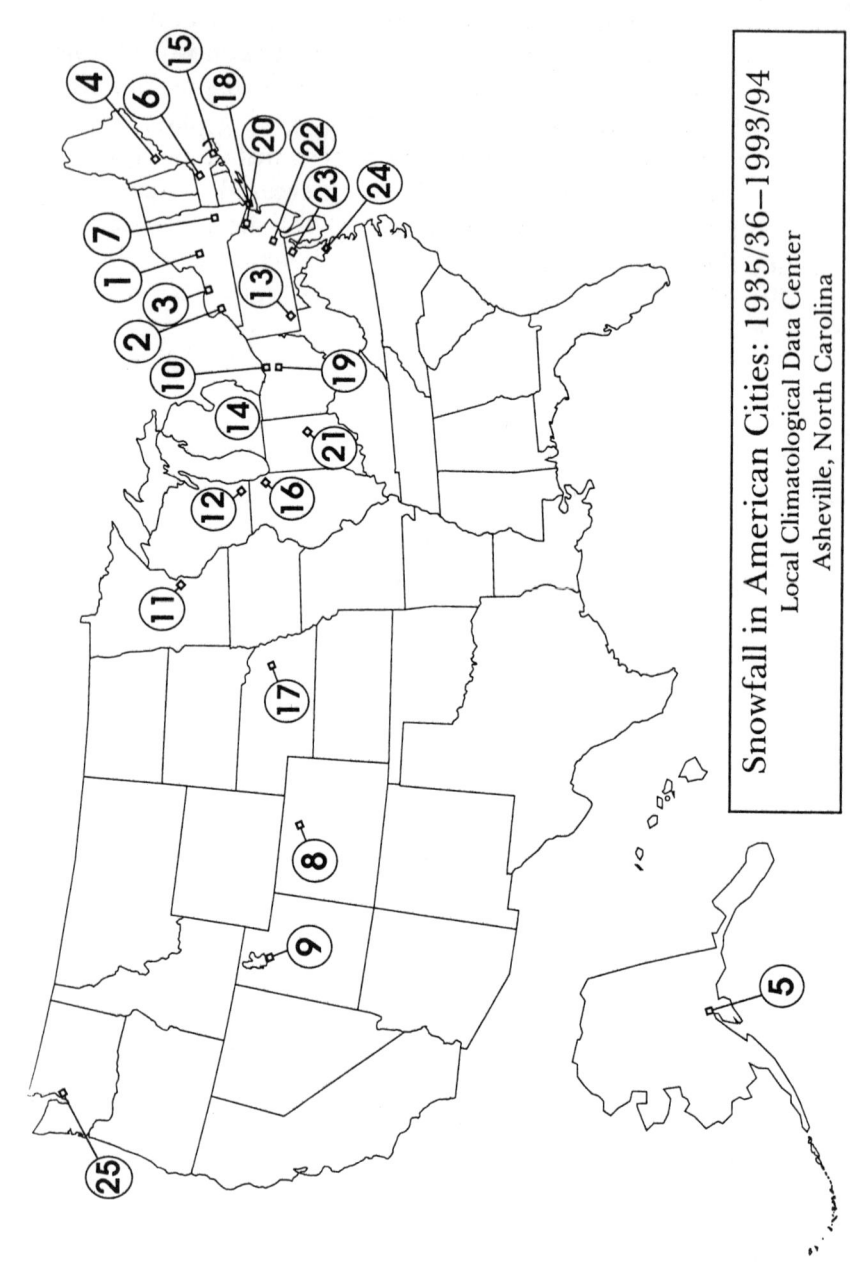

	Cities	1934-94 Mean In inches	Highest Monthly Total In inches	Highest Seasonal Total In inches
1	Syracuse	113.9	72.2 in 1/78*	129 in 92/93*
2	Buffalo	91.0	68.4 in 12/85*	199 in 76/77*
3	Rochester	89.8	64.8 in 2/58*	161.7 in 59/60*
4	Portland, Me.	70.5	62.4 in 1/79*	141.5 in 70/71*
5	Anchorage	69.3	35.2 in 12/78	102.4 in 89/90
6	Worcester	67.8	46.8 in 1/87*	120.1 in 92/93*
7	Albany	63.2	57.5 in 12/69*	112.5 in 70/71*
8	Denver	60.4	31.2 in 12/78	99.3 in 58/59
9	Salt Lake City	57.9	50.3 in 1/93*	110.8 in 73/74*
10	Cleveland	55.4	42.8 in 1/78	100.5 in 81/82*
11	Minneapolis	49.6	46.9 in 11/91*	98.4 in 83/84
12	Milwaukee	46.6	42.0 in 2/74	90.8 in 51/52
13	Pittsburgh	43.1	40.2 in 1/78	70.7 in 69/70
14	Detroit	41.2	34.9 in 12/75	74.0 in 81/82
15	Boston	41.1	41.3 in 2/69*	83.9 in 92/93
16	Chicago	38.2	35.2 in 12/78*	83.7 in 78/79*
17	Omaha	29.9	27.2 in 3/48	58.5 in 47/48
18	New York City	29.2	29.6 in 12/47	63.2 in 47/48*
19	Columbus	27.7	34.4 in 1/78	54.1 in 77/78
20	Newark	26.9	27.4 in 1/78	64.9 in 77/78
21	Indianapolis	22.7	30.6 in 1/78	58.2 in 86/87
22	Philadelphia	20.9	27.6 in 2/79	54.9 in 77/78
23	Baltimore	20.6	33.1 in 2/79	51.8 in 63/64*
24	Washington	16.6	30.6 in 2/79*	37.7 in 78/79
25	Seattle	11.8	45.5 in 1/69*	67.5 in 68/69*

*Record breaking month or season; other monthly and seasonal highs were surpassed before 1935.

INDEX

Abbe, Cleveland, 42–43
Akron, Ohio, 105, 119
Albany, New York, 26, 54, 63, 77, 93, 101, 130, 133, 141–142, 147, 165–166
American City, xviii, 84, 95, 148, 151
American Public Works Association, 110, 142
Annapolis, Maryland, 14
Antarctica, 187
Armstrong, Frank, 42
Asia, 189
Asphalt, 79

Baltimore, Maryland, 15, 23, 47, 54, 85, 118, 142, 154, 183, 185, 188
Barre, Vermont, 118
Barton-Ashman snow-fighting plan, 159–60
Baxter, Samuel, 143, 150
Beame, Abraham, 141
Bergh, Henry, 45–47, 49, 59
Best, Gerald, 32
"Big Snow" of 1836, 25
"Blizzard," first press use of term, 39
"Blizzard Ball" (Buffalo), 187
Boston Gazette, 14, 23
Boston, Massachusetts, xvii, 1, 3–9, 11, 13–14, 21, 23–24, 28–29, 34, 36, 47–50, 55, 58, 63, 73–74, 77, 81, 85, 87–88, 90–92, 102, 121, 131, 139–140, 142, 144–145, 148, 161, 184, 187
Boston News Letter, 6–8
Boston Post, 28

Brennan, James, 68–69
Britannia, 28
Bronx, New York, 61
Brookline, Massachusetts, 170
Brooklyn Bridge, 52
Brooklyn, New York, xiv, 25, 35, 52, 55, 63, 109, 164
Buffalo, New York, viii, xi, 1, 25–26, 29, 35–37, 45, 48, 50, 54, 67, 71–72, 74, 77, 85, 88, 90, 103, 107–108, 112–113, 115, 125–126, 130, 133–134, 140, 142–145, 147, 152, 163, 165–167, 173, 176–178, 181, 185, 187, 192
Burlington, Massachusetts, 172, 174

Cairo, Illinois, 89
Cambridge, Massachusetts, 6, 9, 21, 121
Canada, 22, 44, 57, 93, 113, 147, 162, 176
Carter, Jimmy, 176, 178, 180
Casper, Wyoming, 169
Castle Del Key, 7
Charleston, South Carolina, 8–9, 90
Chattanooga, Tennessee, 89
Cheyenne, Wyoming, 169
Chicago, Illinois, 28, 30, 42–43, 46, 54–55, 74–75, 77, 87–89, 92, 95, 103, 117, 126, 132, 135, 150, 159–160, 171, 173, 177, 184, 186, 190
Chicago Daily Press, 29
Cincinnati, Ohio, 16, 29–30, 42, 78, 90, 118, 169, 178, 184

Cleveland, Grover, 70
Cleveland, Ohio, 80–81, 115, 118, 121, 124, 180
Cleveland Plain Dealer, 80–81
Clinton, DeWitt, 27
Coleman, James S., 51–52, 56, 62–63, 68
Colorado, 94
Columbus, Ohio, 178, 180
Concord, Massachusetts, 130
Connecticut, 11, 94, 180
Cooke, Morris L., 85
Corning, [Mayor of Albany], 166
Cornwallis, 13
Cosmopolitan, 73
Cutler, James G., 76

Daley, Richard, 159, 184
Democrat and Chronicle, 153
Denver, Colorado, 77, 79, 106–107, 115, 136, 171, 186
Des Moines, Iowa, 39, 118, 137, 186
Detroit, Michigan, 30, 43, 75, 77, 94, 115–118, 151, 171
Dewey, Chester, 25
Dickinson, William E., 173
Dr. Shena's Sanitarium, xiv
Dover, New Hampshire, 5
Duff, James B., 123
Duluth, Minnesota, 104
Dunkirk, New York, 169
Dunn, Elias B., 57–58, 64

Earl of Bellmount, 5
East Windsor, Connecticut, 22
Edholm, C. L., 87
Edmonton, Canada, 189–190
El Paso, Texas, 186
Electric sweepers, 70–71
Electric trolleys, 64–67, 70–72
Elevated railroads: in Chicago, 74; in New York, 53, 59, 63
Elmira, New York, 78
England, 4
Erie, Pennsylvania, 130
Estherville Vindicator, 39
Europe, 113, 162, 189

Evening Traveller, 34
Exeter, New Hampshire, 21

Fairbanks, Alaska, 108, 115, 158
Fetherston, J. T., 85–86
Fire engines, 60–61; hand pumpers, 27–28; replaced by steam fire engines, 35
Fisher, E. A., 76
Fisher, Karen, 186
Florida, 177
Funerals, 61, 154

Garfield, Harry, 90
Geneva, New York, 94
Georgia, 57
Gettysburg, Pennsylvania, 23
Grand Rapids, Michigan, 169
"Great Blizzard of 1888," 57–62, 64, 187
"Great Blizzard of 1993," xi, 187–188

Hackney coaches, 16
Haley, [Commissioner of Boston], 143
Harriman, [Governor of New York], 133
Harrisburg, Pennsylvania, 114, 116
Hartford, Connecticut, 11, 14–15, 50, 61, 63, 93, 116, 121, 131, 186
Haskell, Timothy, 20
Helsinki, Finland, 139, 189
Hempstead, Joshua, 10
Holidays, school closings, 50, 61, 131, 139, 144, 160–161, 181; snow holidays, 167, 176, 179, 190
Holyoke, Edward, 10
Hooker, C. D., 115
Howe, William, 13
Hughes, Edward, 154
Hull, John, 4
Hutchison, P. A., 86

Ice-boat clubs, 53, 82–83

Index

Ice storm of 1991, 187
Illinois, 184
Indianapolis, Indiana, 104, 151, 178
International Association of Mayors in Northern Cities, 191
International Conference on the Livable Winter City, 182
Iowa, 117
Iowa Falls, Iowa, 39
Itagaki, Takeshi, 189

Jackson, Mississippi, 186
Jefferson, Thomas, 12
Juneau, Alaska, 89, 168
Jurewiscz, Edward, 145

Kalamazoo, Michigan, 169
Kansas, 48
Kansas City, Kansas, 48
Kansas City, Missouri, 53, 79, 83, 119, 138, 171
Kansas City Times, 79
Kenlon, John, 92
Kennedy, John F., 140
Kenosha, Wisconsin, 104
Key West, Florida, 188
Knox, Henry, 13

Laconia, New Hampshire, 94
Lake Ontario, 72
Lancaster, Pennsylvania, 6, 23
Lapham, Increase A., 42–43
Lee, Richard, 135
Lincoln, Abraham, 29
Lindsay, John V., xiii, 161, 163–166
"Livable winter" programs, 187
Liverpool, England, 86
London, England, xvii, 58, 86
Long Island, New York, 11, 14, 52
Looting, 159
Louisville, Kentucky, xi, 54, 90, 178
Lucia, Frank J., 142–145
Ludlum, David M., ix-x, xx, 2–3, 21, 121, 182, 185
Lynn, Massachusetts, 21

Macon, Georgia, 171
MacSparren, James, 10
Maine, 8, 11
Makowski, Stanley M., 177
Manhattan, New York, 14, 63, 160, 164, 185
Maryland, 37, 177
Massachusetts, 94, 172, 174, 180, 184
Mather, Cotton, 5, 8
Mather, Increase, xi, xvi, 5
Memphis, Tennessee, 70, 89, 124, 148
Michigan, 94
Miller, Edwin A., 111–112
Miller, James E., 121
Milner, Samuel, 186
Milwaukee Sentinel, 33–34
Milwaukee, Wisconsin, 26, 28, 30, 32, 42–43, 51, 54–55, 75, 88, 103, 106, 121, 123, 137, 145, 159, 165, 168
Minneapolis, Minnesota, 53, 77, 88, 105–106, 123–124, 182, 185, 187, 189
Minnesota, 48, 172
Mississippi, 37
Montreal, Canada, 81, 147, 162, 168, 191
Moscow, Russia, 139, 163
Motor plows, 72, 77, 86–89, 92–93
Municipal Journal, xviii, 84
Muskegon, Michigan, 168–169
Myer, A. J., 42–43

Namias, Jerome, 137
Narragansett, Rhode Island, 10
New Bedford, Massachusetts, 23
New Brunswick, New Jersey, vii, 102
New Castle, Pennsylvania, 6
New Haven, Connecticut, 9, 14, 21, 63, 135
New Jersey, vii, 23, 58, 94, 102, 140, 165
New London, Connecticut, 10, 21
New Mexico, 186

New Orleans, Louisiana, 148
New York City, New York, x, xvii-xviii, 5–7, 9, 11–13, 21–25, 27, 30–31, 34–37, 42, 44–49, 52–56, 59–60, 63–64, 67–69, 72, 73–74, 78, 81–82, 84–88, 90–92, 101–102, 105, 108–109, 118–119, 121, 125, 129, 131, 134, 137, 139, 142–146, 148–149, 151, 154, 160–161, 163, 165, 170, 174, 177–178, 185, 187–188, 191
New York Graphic, 51
New York Herald, 25, 59, 95
New York State, xiii-xiv, 8, 22, 25, 53, 57–58, 72, 93–94, 115, 117, 119, 134, 146, 162, 178, 189
New York Sun, 60
New York Times, x, 33–35, 37–38, 43, 46–47, 60, 63, 74, 81–82, 102, 132, 139, 160, 185
New York Weekly Journal, 9
New York World, 60
Newark, New Jersey, 88, 118
Norfolk, Virginia, 14
Norris, Isaac, 6
North American Snow Conference, 158, 169
North Carolina, 12
Northeast Conference on Urban Snow Removal, 142–145, 149–150
Northern Intercity News, 190

O'Connor, James F., 149
Ohio, 22, 123, 134, 180
Omaha, Nebraska, 43, 121–122, 137–138, 171
Omaha World Herald, 138
Oregon, 53
Oswego, New York, 72, 77, 135–136, 154–155, 166–167, 169
Ottawa, Canada, 162
Ozone layer, 158

Paine, H. E., 42
Paris, France, 86
Paxton, J. W., 85

Pennsylvania, 15, 22, 37, 94
Philadelphia, Pennsylvania, 6–7, 9, 11, 22, 24, 26–27, 29, 32, 55, 62, 68, 74–76, 84–88, 92, 102–103, 116, 140, 142–143, 145–146, 149, 158, 178, 183, 185, 188
Pierakos, James L., 181
Pike, John, 5
Pittsburgh, Pennsylvania, 23, 70, 73, 77, 85, 90, 94, 102, 119, 121, 123, 132, 142, 144–145
Pittsburgh Post, 95
Pittsburgh Post Gazette, 71
Plymouth, Massachusetts, 2–3
Portland, Maine, 34, 38, 43, 100–101, 131, 158, 161–162, 182–183, 187, 190
Portland, Oregon, 53, 108, 115, 122, 124, 135
Portland Press Herald, 161
Portsmouth, New Hampshire, 6, 12
Providence, Rhode Island, 21, 93, 116, 180
Public Works, 93, 168, 173–174

Quebec City, Canada, 82, 147
Queens, New York, xiii, 164

Randolph Weekly Wanderer, 22
Rapid City, Iowa, 169
Reston, James, 164
Retsof, New York, 118
Rhode Island, 180
Rhodes, James A., 180
Riis, Jacob, 69
Robbins, Thomas, 22
Rochester, New York, vii, xi, xiv-xv, xviii, 20, 25, 29, 32–33, 35–37, 43–45, 48, 50, 52–53, 56, 67, 71–72, 74–76, 81–82, 85, 87, 89–90, 100, 105, 110, 112, 119, 126, 130–131, 133–136, 138, 147, 152, 165–167, 175, 179, 181, 187
Rockefeller, Nelson, 164, 166
Rogers, William C., 182

Index

Rome, New York, 93, 151
Rooney, John F., 168–169
Roosevelt, Theodore, 69
Ruggles, George W., 71
Ruggles plow, 71
Russell, Leonard W., 100
Rutland, Vermont, 21

Sacramento, California, 32
Sagadahoc, Maine, 2
St. Louis, Missouri, 30, 38, 48, 54, 83, 178
St. Paul, Minnesota, 39, 48, 51, 77, 106, 124, 135, 182, 185
Salem, Massachusetts, 10, 21
Salem, Oregon, 108
Salt, 49, 86, 99, 115–119, 125, 132–135, 144, 160; opposition to use, 171–75
Salt Lake City, Utah, 187
San Antonio, Texas, 185
San Francisco, California, 43
Sapporo, Japan, 189–191
Savannah, Georgia, 9, 12
Schenectady, New York, 115
Scher, (City Manager of Rochester, New York), 153–154
Scientific American, 73, 84
Scrantom, Edwin, 20
Scranton, Pennsylvania, 22, 85
Seattle Post-Intelligencer, 123
Seattle, Washington, 88, 122–123, 170
Sedita (Mayor of Buffalo, New York), 152
Seneca Falls, New York, 35
Severance, Paul R., 140
Sewall, Samuel, 3, 5
Shenyang, China, 190
Shobie, Illinois, 185
Skaters, 23, 53, 114, 147
Skiing, 85, 113, 141, 147, 162, 167, 176, 179, 190
Sleighs, 6–7, 11, 21, 25, 41, 45, 54
Smithsonian Institution, 19, 37, 41–42
Snow conferences, 84–86; (*see also* Northeast Conference on Urban Snow Removal; North American Snow Conference)
Snow fences, 104
Snow melters, 86, 146, 170
Snow plows, 24–25, 32–35, 41, 72–73
Snow rollers, 94, 163
Snowmobiles, 154, 165, 167
Society for the Prevention of Pauperism, 27
Solomon, x
South Amboy, Massachusetts, 24
Spokane, Washington, 88, 122
Springfield, Illinois, 29
Springfield, Massachusetts, 115
Stockholm, Sweden, 162
Street Cleaning Practice, 110, 135
Strong, George Templeton, 30–31, 38, 46, 191
Subways, 63, 101
Syracuse Herald, 96–97
Syracuse, New York, viii, 25, 35, 37, 50, 72, 77, 81–82, 87, 90, 95, 97, 100, 112, 125, 130–131, 133–134, 138, 140, 142, 147, 154–155, 158, 166–167, 180, 182, 187–188
Syracuse Post-Standard, 134

Taft, William, 78
Technological Manual on Winter Road Management, 191
Telegraph lines, 19, 29, 39, 41, 51, 55, 59, 74, 80
Temmerman, John, 119, 133, 135
Thermometers, 8, 15
Ticonderoga, New York, 13
Tire chains, 121
Tiros I, 136
Tiros II, 137
Tiros satellite program, 183
Tomaso, Norway, 191
Tonawanda, New York, 152
Toronto, Canada, 93, 104
Traverse City, Michigan, 130
Troy, New York, 104
Truman, Harry S., 122

Utica, New York, 22, 25, 53, 77, 93, 142, 147

Vanguard II, 136
Vermont, 20, 22, 72
Very, E. D., 86
Virginia, 4–5, 8, 12, 15, 37

Wagner, Robert, 131, 140–141, 151
Walrath, John H., 96
Waring, George E., 69, 73, 78
Warner, Carl D., 118
Washington, D.C., 19, 37–38, 43–44, 50, 54–55, 57, 70, 74–75, 77, 85, 90, 93, 133, 140, 142–145, 154, 161, 170, 178, 181, 183, 187–188
Washington, George, 12–13
Watertown, New York, 169, 178
Watson, John F., xix
Watts, James, 11
Weather Bureau, 136–137, 186, 188
Weather watchers, 6, 15, 19, 136

Weatherwise, ix, 120, 131, 136, 183, 186
Webster, Noah, 3, 15, 21
Western Union, 29
"White Hurricane", 1, 188
Wicks, C. W., xiv
Wilkes-Barre, Pennsylvania, 130, 132
Williams, [Captain of New York City Police Department], 51
Williamsburg, Virginia, 12
Williamsport, Pennsylvania, 151
Wilson, Woodrow, 91
Winter carnival: in Albany, 54; in Rochester, 82; in St. Paul, 185
Winthrop, John, 4–5
Winthrop's Journal, 3
Wisconsin, 26, 34, 159, 185
Wood, William, 2–3
Woodbury, John M., 78
Worcester, Massachusetts, 14, 21, 24, 73, 93, 100, 130, 187

Xanten, William, 144, 149

"Year without a summer" (1816), 22